POLYMER SURFACES AND INTERFACES

Acid-Base Interactions and Adhesion in Polymer-Metal Systems

ABOUT AAP RESEARCH NOTES ON POLYMER ENGINEERING SCIENCE AND TECHNOLOGY

The AAP Research Notes on Polymer Engineering Science and Technology reports on research development in different fields for academic institutes and industrial sectors interested in polymer engineering science and technology. The main objective of this series is to report research progress in this rapidly growing field.

POLYMER SURFACES AND INTERFACES

Acid-Base Interactions and Adhesion in Polymer-Metal Systems

Irina A. Starostina, DSc, Oleg V. Stoyanov, DSc, and Rustam Ya. Deberdeev, DSc

Apple Academic Press Inc.
3333 Mistwell Crescent
Oakville, ON L6L 0A2
Canada

Apple Academic Press Inc.
9 Spinnaker Way
Waretown, NJ 08758
USA

First issued in paperback 2021

Exclusive worldwide distribution by CRC Press, a member of Taylor & Francis Group
No claim to original U.S. Government works

ISBN 13: 978-1-77463-325-0 (pbk)
ISBN 13: 978-1-926895-99-4 (hbk)

Library of Congress Control Number: 2014937679

Library and Archives Canada Cataloguing in Publication

Starostina, Irina A., author
Polymer surfaces and interfaces: acid-base interactions and adhesion in polymer-metal systems/Irina A. Starostina, DSc, Oleg V. Stoyanov, DSc, and Rustam Ya. Deberdeev, DSc.

Includes bibliographical references and index.
ISBN 978-1-926895-99-4 (bound)
1. Polymers–Surfaces. 2. Interfaces (Physical sciences). 3. Composite materials. 4. Acids–Basicity. 5. Metals. 6. Adhesives. 7. Adhesive joints. I. Stoyanov, Oleg V., author II. Deberdeev, R. Ya, author III. Title. IV.
Series: AAP research notes on polymer engineering science and technology series

QD381.9.S97S73 2014 620.1'920429 C2014-902427-4

Apple Academic Press also publishes its books in a variety of electronic formats. Some content that appears in print may not be available in electronic format. For information about Apple Academic Press products, visit our website at **www.appleacademicpress.com** and the CRC Press website at **www.crcpress.com**

ABOUT THE AUTHORS

Irina A. Starostina, DSc

Irina A. Starostina, DSc, is Professor of the Chair of Physics at Kazan State Technological University in Russia. She has dealt with physicochemistry of polymer adhesion for over 20 years.

Oleg V. Stoyanov, DSc

Oleg V. Stoyanov, DSc, is Head of the Chair of Industrial Safety and Professor of the Chair of Technology of Processing Plastic and Composite Materials (TPPCM) at Kazan State Technological University in Russia. He is a world-renowned scientist in the field of chemistry and the physics of oligomers, polymers, composites, and nanocomposites.

Rustam Ya. Deberdeev, DSc

Prof. Rustam Ya. Deberdeev (DSc) is the honored worker of science and the laureate of State Prizes of the Republic of Tatarstan and the Russian Federation. His scientific interests involve synthesis, technology, and processing of polymers and composite materials.

BOOKS IN THE AAP RESEARCH NOTES ON POLYMER ENGINEERING SCIENCE AND TECHNOLOGY SERIES

- **Functional Polymer Blends and Nanocomposites: A Practical Engineering Approach**
  Editors: Gennady E. Zaikov, DSc, Liliya I. Bazylak, PhD, and A. K. Haghi, PhD

- **Polymer Surfaces and Interfaces: Acid-Base Interactions and Adhesion in Polymer-Metal Systems**
  Irina A. Starostina, DSc, Oleg V. Stoyanov, DSc, and Rustam Ya. Deberdeev, DSc

- **Key Technologies in Polymer Chemistry**
  Editors: Nikolay D. Morozkin, DSc, Vadim P. Zakharov, DSc, and Gennady E. Zaikov, DSc

- **Polymers and Polymeric Composites: Properties, Optimization, and Applications**
  Editors: Liliya I. Bazylak, PhD, Gennady E. Zaikov, DSc, and A. K. Haghi, PhD

CONTENTS

CONTENTS

LIST OF ABBREVIATIONS

TDQ	2,2,4-Trimethyl-1,2-dihydroquinoline
BR	Butadiene rubber
IIR	Butyl rubber
CIIR	Chlorobutyl rubber
CS	Curing system
DETA	Diethylenetriamine
DMFA	Dimethylformamide
DMSO	Dimethylsulfoxide
DPP	Diphenylolpropane
ES	Ethyl silicate
EPDM	Ethylene propylene diene monomer
EBA	Ethylene-butyl acrylate copolymer
EEA	Ethylene-ethyl acrylate copolymer
EEA terpolymer	Ethylene-ethyl acrylate-acrylic acid terpolymer
EVA	Ethylene-vinyl acetate copolymer
EVA terpolymer	Ethylene-vinyl acetate-maleic anhydride terpolymer
HSAB	Hard Soft Acid Base
HDPE	High density polyethylene
IR	Isoprene rubber
LDPE	Low density polyethylene
MPDM	m-Phenylenedimaleimide
OPD	o-Phenylenediamine
PDB	p-Dinitrobenzene
PQD	p-Quinone dioxime
PAT	Polyamine grade
PC	Polycarbonate
PE	Polyethylene
PEPA	Polyethylene polyamine
PET	Polyethylene terephthalate
PI	Polyisocyanate

PPR	Polymeric petroleum resin
PMMA	Polymethyl methacrylate
PTMG	Polyoxytetramethyleneglycol
PP	Polypropylene
PPC	Polypropylene carbonate
PS	Polystyrene
PTFE	Polytetrafluoroethylene
PVC	Polyvinylchloride
PAA	Primary aromatic amine
SEM	Scanning electron microscopy
SBR	Styrene butadiene rubber
SFE	Surface free energy
vOCG	van-Oss-Chaudhury-Good method
VA	Vinyl acetate
ZDEC	Zinc diethyldithiocarbamate

LIST OF SYMBOLS

D	Acidity parameter
ΔD	Relative acidity parameter
θ	Contact angle
γ	Surface free energy
γ^d	Dispersive SFE component
γ^{ab}	Acid-base SFE component
W_a	Thermodynamic work of adhesion
W_a^d	Van der Waals component of the thermodynamic work of adhesion
W_a^{AB}	Acid-base component of the thermodynamic work of adhesion
ΔH_{sl}^{AB}	Enthalpy of the acid-base interaction between two materials in contact

PREFACE

The papers on adhesion available to modern reader are permanently at a high level. However, the acid-base approach explicated in this monograph regarding the enhancement of adhesive interaction has neither been considered in detail nor applied to a broad range of metal-polymer compounds. The reviews available, for example, written by F.M. Fowkes and Della Volpe [1, 2], or individual monograph sections [3] cover at most only a dozen polymers that give no possibility to generalize results and predict the adhesive ability of materials on the basis of their acid-base properties. At the same time, studies having been carried out by authors since 1993, in co-authorship with doctoral students and doctoral candidates E.V. Burdova, R.K. Khairullin, V.Ya. Kustovsky, and A.V. Chernov, who earned Ph.D. (C.Sc.) degrees in Engineering, allowed experimental data on surface properties of more than 150 polymers, such as carbocatenary and heterochain polymers, copolymers and their blends, and as well as different epoxy and rubber compositions used in adhesive joints, to be gathered and systematized.

This present monograph involves the analysis of all up-to-date techniques used for determination of acid-base properties in view of their applicability to examination of solid organic and inorganic surfaces, and places emphasis on the most promising methods. Also, the adhesive ability of metal-polymer systems based on epoxy compositions, polyolefins, and rubbers was studied as a function of absolute difference in acid-base properties of adhesive and adherend, and the possibility to predict adhesive interaction on this basis was experimentally verified.

We hope that this book will interest not only specialists who design, produce and utilize adhesive joints but also students and doctoral students who are focusing on adhesive interaction improvement.

We express a special gratitude to Professor R.M. Garipov for his help and attention paid for the revision of this monograph.

INTRODUCTION

Creating polymeric materials with greater strength characteristics when in contact with metals is the most important problem when adhesive joints are designed. But usually, the adhesive ability of polymer-metal systems is controlled in most cases by optimizing formulation and process parameters that do not allow solving the problem scientifically in whole.

More attention has been recently paid to acid-base interactions in order to study their role in interfacial bond formation. The best interaction is achieved if one of the adhered materials preferably has the properties of a Lewis acid, whereas the other one has the properties of a Lewis base. The acid-base theory for adhesive systems has become much more developed and commonly recognized in the recent years. Today, the theory is at the stage of generating, gathering and analyzing experimental data. However, the common approach, which could allow estimation of potentially possible interfacial interaction, prediction and control of this interaction, is still undeveloped. There are odd bits of existing experimental data that are virtually lacking in the case of composite materials. Thus, the determination of the acid-base properties of polymer surface as well as the relationship between these properties and interfacial interaction in adhesive joint is of particular scientific interest. It is worthwhile noting that estimation of acid-base characteristics of the prepared solid polymer surfaces and different low molecular weight additives, such as, fillers, plasticizers, promoters, etc., does not seem today as a trivial task. Most existing techniques for estimating the aforementioned properties find limited application to polymers. First of all, a reasoned choice of correct quantitative characteristic of the acid-base properties of adhesive joint components is needed. Therefore, this monograph contains a comparative analysis of up-to-date techniques for estimating surface-energy and acid-base characteristics of solid smooth surfaces of polymers and metals. The Berger method and the van-Oss-Chaudhury-Good (vOCG) method were used as priority wetting techniques. Today, the vOCG method is used in foreign adhesion practice as the most in demand and, at the same time, questionable enough, being therefore our special interest. On the other hand, our examinations once and again showed that the Berger method ensures practical available information for producing adhesive joints with greater strength characteristics.

The aim of this monograph is to show an important role that acid-base interactions play in establishing interfacial adhesive-adherend contact, and to outline practical recommendations regarding parameters of quantitative estimation of acid-base surface properties, which implies the relationship with adhesive ability in polymer-metal systems.

The information offered to the reader is a first extensive experimental verification of the acid-base theory statements in practice, under laboratory and industrial conditions, applied to a wide range of polymer-based compositions used as coatings and adhesives of different purpose, which makes systematization and statistical analysis of results possible. We obtained experimental data for thermodynamic and acid-base properties of about 200 organic and inorganic surfaces, which find wide practical application. These results may be used as a reference source to predict the adhesive ability of different coating systems. The possibility to predict adhesive interaction of adhesive with adherend, taking into account the absolute difference in their acidity and basicity, was verified by us experimentally. All experimental results were implemented in practice. The optimized adhesive formulations were proposed and successfully passed laboratory tests at different Russian enterprises.

— Irina A. Starostina, DSc, Oleg V. Stoyanov, DSc,
and Rustam Ya. Deberdeev, DSc

CHAPTER 1

BASIC THEORETICAL STATEMENTS OF ACID-BASE INTERACTION

CONTENTS

1.1 HISTORICAL BACKGROUND

An acid-base (donor-acceptor) interaction is a special kind of interaction involving one particle as a donor, and the other one as an acceptor of electron pair.

An acceptor is a molecular (atomic) system having unoccupied levels (orbitals) and a positive electron affinity. A system having a lone electron pair is referred to as a donor. The electron pair becomes common for donor and acceptor when a bond is formed [4]. The forces of this nature are of our main interest.

The basics of modern theory of protolytic acid-base equilibrium, usually associated with Brønsted who developed it quantitatively, were stated in 1923.

According to Brønsted [5], an acid is a proton donor (a reagent which donates a proton) and a base is a proton acceptor (a reagent which receives a proton). These reagents may present in the form of either molecules or ions. The reaction of proton transfer itself is referred to as protolytic. As protolytic reaction is reversible, with proton transfer occurring also in the reverse process, the reaction products act also as an acid and a base relative to each other. The acid-base properties of substances depend on the thermodynamics of protolytic reactions.

It is worthwhile noting that the Brønsted reaction is thermodynamic and concerns only the equilibrium of protolytic reactions in diluted solutions, rather than their kinetics. Many experimental data are in agreement with this theory. However, there are a number of discrepancies caused by its approximate nature concerned with the electrostatic effect of interionic interaction only, rather than a special character of chemical interaction of a solvent with an acid or a base.

At about that time Lewis [6] proposed another approach to the concepts of acids and bases, based on electronic structure of a molecule. According to Lewis, the acidic electron-acceptor properties depend on the structure of reacting molecules only, irrespectively of the presence of any specific element, such as hydrogen atom, which can brake away as a proton.

A distinctive feature of acids and bases according to the electronic theory is their mutual neutralization occurring through the formation of a covalent bond between the atom of base, having a lone electron pair, and the atom of acid, having an electron shell where the electron pair is included. The electron pair becomes common for the corresponding atoms of the product of neutralization. Thereby, the acid-base process differs from the redox one which implies a complete transfer of a single or more electrons from a reducer to an oxidizer.

According to Lewis, a proton itself as a particle, which easily receives an electron pair refers to strong acids. According to the electronic theory, Brønsted acids are formally considered as products of neutralization of a proton by bases. The characteristic of Lewis acids and bases is their specific interaction. Therefore, it is impossible to arrange them in a row by strength, since this arrangement will depend on substances used as reference standards. Due to the aforementioned specific nature of the interaction of Lewis acids and bases as well as quantum-chemical effects to be hardly taken into account, the electronic theory does not seem so clear and quantitative compared with the theory of protolytic equilibrium. However, not only does it allow expanding the range of substances showing acid-base properties but also more extensively reveals the origin of these properties stipulated by structure of substances.

Both approaches are interrelated because a proton itself, like other Lewis acids, is characterized by a strong affinity to electron pair.

Thus, a Lewis acid is an electron acceptor, and a Lewis base is an electron donor, whereas a Brønsted acid is a proton donor, and a Brønsted base is a proton acceptor. In up-to-date literature, it is usual to indicate which one, a Lewis or Brønsted acid is meant to avoid any confusion.

Other acid-base theories are similar, and rather complementary, to those mentioned above. For instance, A.I. Shatenstein [7] offered the initial (hydrogen bond formation) and final (complete proton transfer) stages of acid-base interaction; the reaction may involve the first stage only. Instead of proton acids, Shatenstein suggested substances, which do not deprotonate but exist in equilibrium with bases as proton acceptors, to be referred to as acid-like aprotic substances. The acid concept proposed by N.A. Izmaylov [8] is about dissociation of acid dissociation products rather than about acids themselves. Thus, adducts of solvent with acid, solvated ions, and ionic pairs may exist in solution together in equilibrium. According to Izmaylov, an acid is a substance containing hydrogen and taking part in acid-base interaction as a proton donor, whereas a base is a substance participating in acid-base interaction as a proton acceptor, with hydrogen bond formation being the necessary characteristic of acid-base interaction.

Further development of the theory of acid-base interactions resulted in the solvent system (ionotropy) theory [9] and the Lux-Flood oxidotropic concept [10]. The solvent system theory considers the ability of a solvent to dissociate, followed by heterolytic bond cleavage, as a main solvent characteristic. Released ion is bonded with solvent molecule to form anions and cations in solvent during dissociation. Solvent dissociation is considered as ionotropy, the process of ion transfer between the reactants. Solvo-acids are substances, which increase solvent ion concentration after dissolution and dissociation, and solvo-bases are substances, which increase anion concentration.

According to the oxidotropic concept, the properties of acids and bases are described basing on the O^{2-} oxide-ion behavior.

As a result of developing the Lewis acid-base theory, the really elegant Hard Soft Acid Base (HSAB) theory was created [11]. The HSAB theory involves both kinetic and thermodynamic aspects. This should be understood to mean that in case of combining hard bases and hard acids as well as soft bases and soft acids, the interaction occurs at relatively greater velocity to give relatively more stable products.

In Ref. [12], ligand atoms and metal ions are divided into soft and hard bases and acids according to their electronegativity and polarizability. Hard acids and bases are strongly polar but less polarizable molecular systems and their fragments, whereas soft acids and bases are the reverse.

In 1939, M.I. Usanovich proposed the most general acid-base theory, which involves other theories. According to Usanovich [13], acids are substances, which donate cations or receive anions (electrons); bases are substances which donate anions (electrons) and receive cations. According to this theory, the quantitative method for estimating acid or base degree as a function of ionic potential z/r (where z – the ion charge number, r – the ion radius) was proposed. If a central ion of a substance is positively charged, the acidity increases as the charge raises and radius decreases. If the ion is negatively charged, the acidity decreases, but basicity increases as the potential rises.

By generalizing a great number of acid-base theories, the main similarity between them may be found. All theories define an acid as a substance which donates the positively charged species (hydrogen cation, solvent cation, etc.) or receives the negatively charged species (oxide ion, electron pair, etc.). A base is defined as a substance, which donates the negatively charged species (electron pair, oxide ion, solvent anion, etc.) and receives the positively charged species (hydrogen cation, etc.).

The definitions given may be generalized as follows: the acidity is a positive nature of chemical species with concentration which decreases in the reaction with base; whereas the basicity is a negative nature of chemical species with concentration which decreases in the reaction with acid [14].

Millikan defined acid-base interactions as complexes for charge transfer and extracted two contributions, namely electrostatic and covalent, to the energy of these interactions [15].

1.2 ACID-BASE ABILITY OF SUBSTANCES

By its acid-base ability, an object may be referred to one of three types [7, 16–18]:

(a) Electron acceptors, or proton donors, that is, acids. These may include, for example, halogenated molecules, including polymers like polyvinylchloride; polyvinylidene fluoride; chlorinated polyethylene; molecules with polarized double bonds in which the more positive atom is responsible for acid properties of a substance (CO_2, SO_2); halides with unsaturated coordination bonds ($TiCl_4$, SbF_5); cations as central atoms of complex compounds (Fe^{3+}, CO_2^+); molecules with unoccupied eight-electron configuration (BF_3, SO_3); and as well as the SiO_2, Fe_3O_4, and Fe_2O_3 substrate surfaces.

(b) Electron donors, or proton acceptors, that is, bases. These include molecules with stretched electron orbitals (NH_3,); anions as ligands of complex compounds; esters; ketones; ethers; aromatic compounds, and polymers such as polymethyl methacrylate, polystyrene, ethylene-vinyl acetate copolymers, polycarbonates, polyimides, as well as surfaces of substrates like $CaCO_3$, hydrated Fe_2O_3, Al_2O_3, and hydroxides or iron oxide.

(c) Electron acceptors (or proton donors) and electron donors (or proton acceptors) simultaneously, that is, bipolar or amphipathic substances showing the properties of both acid and base. These, for example, are the amide, amine, and hydroxyl-containing polyamides and polyvinyl alcohol, as well as the α-Al_2O_3, AlOOH, and $Al(OH)_3$ substrate surfaces.

There are many substances which are bases showing a slight acidity, or acids showing a slight basicity. Such liquids virtually do not form specific self-associations, and their cohesion energy and surface tension are the functions of their dispersion interactions only [19]. Van Oss et al. introduced the term "monopolar" to characterize substances showing either acid or base properties, or "bipolar" for substances showing both properties [20].

Fowkes established the four classes of substances able to form acid-base bonds [21] and mentioned that any substance of these classes is often referred to as polar by mistake:

1. Strongly hydrogen-bonded substances with a high dielectric permittivity, such as water, glycerol, formamide. These liquids show a low oil solubility due to a strong self-association;
2. Self-associated substances with a high dielectric permittivity and low oil solubility which do not form hydrogen bonds (acetonitrile, dimethyl formamide, nitrobenzene);
3. Dipole substances with a low dielectric permittivity, high oil solubility, and slight self-association ability, such as ethers and esters, ketones, tertiary amines, chloroform, and methylene chloride. These all can form hydrogen bonds with other substances;
4. Substances with a slight or zero dipole moment, low dielectric permittivity, high oil solubility, zero self-association, able to form specific complexes with other substances (benzene, hexamethylene tetramine). Some metal alkyls, in spite of little dipole moments, actively form acid-base complexes.

1.3 CONTRIBUTION OF ACID-BASE INTERACTION TO ADHESION

As mentioned above, the acid-base interaction of adhesive with substrate may play a key role in the formation of adhesive bonds acting across the interface. Obviously, intermolecular acid-base interactions affect solubility, adsorption, and adhesion of polymers to other materials to a great extent, that makes quantitative characterization of acid-base properties of the most frequently used solvents, polymers, and inorganic films and substrates necessary. Today, the leading scientists majoring in adhesion successfully use the acid-base approach for improving adhesive characteristics of substances applicable in different fields, from corrosion protection to stomatology.

Today, there are four main adhesion theories: mechanical, electrical, diffusive and adsorptive. But those researchers who explain the adhesive interaction within a framework of a single theory only are not right a priori. It seems more reasonable to point out basic mechanisms of adhesive joint formation rather than individual theories, and to evaluate their roles played for obtaining the final product. However, taking into account complicated and multifactor nature of the adhesion phenomenon, this approach does not allow the problem to be solved scientifically in whole, and the adhesive ability of polymer-metal systems is still controlled in practice mostly by optimizing formulation and process parameters.

Not for detraction of importance from other theoretical approaches, but for evidence of a significant role the acid-base bonding plays in interfacial interaction of polymers with metals, we therefore approached to the problem extensively, and the attempt was made to minimize the influence of other forces, that is, mechanical, diffusive, etc. by creating special conditions.

It should be noted first that a role the mechanical factors, such as roughness, play in the adhesive interaction may be realized due to special metal treatment used to prepare a given layer relief (etching, bead- and abrasive-blasting, etc.). In present study, according to examination carried out with scanning probe microscope during routine treatment, that is, grinding of metal substrates, it was found that roughnesses have sizes on microlevel.

According to Kinlock et al., the microroughness of substrates are thought to be correlated weakly enough with durability of their adhesive joints [3].

We found that epoxy coatings of the same composition applied on steel of different purity grades (7, 8, and 10) show the same resistance to cathodic peeling upon polarization. Similar results were obtained for polyolefin coatings based on ethylene-vinyl acetate (EVA) copolymers. It was also found that metal substrates having more developed mircorelief (duraluminum and copper) do not provide better adhesive interaction with epoxy coatings of similar composition, compared with metals having less developed relief (steel and brass) (*see* Table 1.1).

TABLE 1.1 Cathodic disbandment of adhesive joint from metals.

Adhesive joint	Coating – St3 steel	Coating – L62 brass	Coating – copper	Coating – D16AM duraluminum
Roughness size, μm	1	1	3–4	3
d_{def}*, mm	5	8	17	17

* The diameter of defect area resulted from cathodic peeling.

Thus, in case of adhesive joint characterized by substrate roughness on microlevel, mechanical interlocking does not noticeably affect interfacial interaction with adhesive. Moreover, mechanical interlocking seems to have zero effect on interaction processes at polymer-metal interface when water, plasticizer, or inhibitor molecules are added, etc. [3]. The increase in adhesive force due to the increased surface roughness may be usually caused by removal of weak interfacial layers during interface treatment, better spreading conditions, and enhancement of energy dissipation mechanism.

Ultimately, it should be indicated that for producing compounds with greater adhesive properties it is usual to deal with substrates with relatively rough surface relief, provided that the oxide layer has suitable mechanical characteristics but adhesive, or primer, can penetrate with pores of any porous oxide. Nevertheless, the presence of mechanical interlocking in the interface area does not critically contribute to the increase in adhesive joint strength.

When a role of diffusive processes in adhesive contact is considered, the known fact that those are of importance for only polymer-polymer systems should be mentioned.

The electrical adhesion theory, from a physical point of view, is a constituent part of the adsorptive theory, since the nature of phenomena under consideration involves donor-acceptor interactions at the interface. The theory's founder B.V. Deryagin compares the formation of a double electrical layer on adhesive joint to donor-acceptor bonding of the two huge molecules [22].

Recently, it is usual to believe that in most cases the adsorptive theory is applicable, since molecular forces in surface layers of adhesive and substrate, significantly affecting interfacial molecular contact, often serve as a primary prerequisite for adhesive interaction. Other theories may be used for determining contribution of additional factors to adhesive interaction. Investigations carried out over two last decades of twentieth century within the framework of the adsorptive adhesion theory by many researchers, including authors of this monograph, confirm an important role the acid-base interactions play in establishing adhesive contact.

The following chapters involve the results we obtained to reveal the character of adhesive interaction in different metal-polymer systems, the evidence of the prevailing role of acid-base interactions in adhesive contact, and possibilities to estimate acid-base properties of solid organic and inorganic surfaces.

KEYWORDS

- **electronegativity**
- **polarizability**
- **protolytic acid-base equilibrium**
- **proton acceptors**
- **proton donors**

CHAPTER 2

THERMODYNAMICS OF SURFACE PHENOMENA IN POLYMER-METAL ADHESIVE JOINTS

CONTENTS

2.1 THERMODYNAMIC WORK OF ADHESION AND SURFACE FREE ENERGY

As is known, the state of thermodynamic system is characterized by thermodynamic potentials U (internal energy), F (free energy), G (Gibbs energy), H (enthalpy). In principle, any of these potentials, under corresponding conditions, is suitable for characterizing the energy of surface acid-base interactions since the work done by thermodynamic system in any process is accompanied by the loss of thermodynamic potential, which meets process conditions.

A negative value of the corresponding differentials $dU < 0$, etc., serves as a criterion of process irreversibility and the possibility to proceed at certain constant parameters. In these cases, the values of thermodynamic potentials decrease and, at the minimum values, the process comes to equilibrium.

Gibbs interpreted the interface as the idealized surface with zero thickness and called it the separating surface. A real interface has a surface layer with finite thickness not exceeding 1 nm, within which thermodynamic parameters (concentration, pressure, temperature, etc.) alter abruptly. Thermodynamic parameters of each phase beyond the separating surface remain unchanged. The interface accumulates the excess of uncompensated energy. This excess per unit surface constitutes the unit surface free energy which is not a specific form of energy but is stipulated by the position of molecules itself at the interface. From energetic point of view, the surface free energy (SFE) is defined as the work of isothermal reversible change in surface layer area per 1 m^2. The SFE may be considered as the work of transferring molecules from bulk onto surface of a solid. The stronger intermolecular forces in the solid, the greater its SFE at the border with gaseous phase [23].

The SFE of objects is one of the most fundamental by its sense and importance parameters, to be a function of the response of criteria and conditions of adhesive joint formation, serving as the physical-chemical equivalent for characteristics of adhesive joint resistance to external action.

The SFE is a partial derivative of any thermodynamic potential with respect to interfacial area at the corresponding constant parameters [23].

The SFE is most frequently expressed in terms of the Gibbs energy derivative since holding T and p constant is experimentally feasible.

The adhesion is related to surface phenomena, which occur spontaneously and result in the decreased surface energy (the Gibbs energy). The spontaneous process condition may be expressed as:

$$\Delta G_{\mathrm{A}} < 0,\ G_{\mathrm{A}} = G/A = \gamma. \tag{2.1}$$

For equilibrium reversible process, the adhesion is estimated with the work W_{a} to be done for reversible dividing a unit surface area of phases in contact. Thermodynamic work of adhesion at the border of two interacting phases is one of the most important criteria of adhesive joint formation. The W_{a} is done against intermolecular forces of electrostatic nature (in this case, attractive forces).

As a result of adhesive joint formation, the unit SFE decreases by the value, which characterizes the work of adhesion as the work of reversible adhesive division of two phases marked as 1 and 2:

$$W_{\mathrm{a}} = -\Delta G_{\mathrm{A}} = \gamma_1 + \gamma_2 - \gamma_{12} \tag{2.2}$$

Henceforth, it can be seen that the W_{a} rises as the interfacial energy γ_{12}, that is, the difference in polarity between phases, decreases, and may be stepped up due to adsorption of surface-active additives at the interface.

In case of a liquid ℓ adhered to a solid s, if the interfacial energy is equal to $\gamma_{\ell s}$ due to adhesion, it becomes equal to $(\gamma_\ell + \gamma_s)$ after the adhesive interaction has been overcome. The equilibrium work of adhesion is equal to difference between the final and starting values of surface energies and may be expressed as:

$$W_{\mathrm{a}} = (\gamma_\ell + \gamma_s) - \gamma_{\ell s}. \tag{2.3}$$

The surface energy of solids γ_s, incorporated in this equation, and the interfacial energy γ_{ls} are hardly determined experimentally, while it can be done easily enough regarding the surface tension of liquid at the border with gaseous phase γ_ℓ. Therefore, the W_{a} is determined by estimating the shape of a drop of liquid on a solid surface upon contact wetting. When a drop of liquid is applied onto a solid surface, a certain finite time is required for drop shape to come to equilibrium state. The equilibrium state of the drop is defined by simultaneous interaction of three surface tensions γ_s, γ_ℓ, and $\gamma_{s\ell}$ (Fig. 2.1). The surface energy γ_ℓ acts at a certain angle θ, which is referred to as the contact angle.

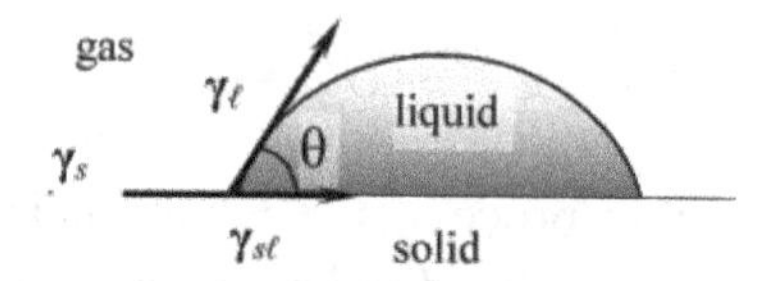

FIGURE 2.1 Determination of the equilibrium contact angle for a drop of liquid on a solid surface.

The contact angle is associated with the surface tension at the border of three phases in the known Young equation [24]:

$$\gamma_s = \gamma_{s\ell} + \gamma_\ell \cos\theta. \tag{2.4}$$

From the Eq. (2.4) we get:

$$\gamma_\ell \cos\theta = \gamma_s - \gamma_{s\ell} \tag{2.5}$$

By inserting the Eq. (2.5) in the Eq. (2.3) the so-called Young-Dupré equation is obtained:

$$W_a = \gamma_\ell(1+\cos\theta). \tag{2.6}$$

Thus, the W_a of a liquid on the surface of a solid is calculated from the Young-Dupré equation by measuring the surface tension of liquid and the equilibrium contact angle. In case of complete wetting this equation cannot be used.

The SFE and thermodynamic work of adhesion are the most important characteristics used for characterization of adhesive interactions. The work of adhesion is related to the SFE of both single surface and adhesive joint in whole and responsible for the potential interfacial interaction ability of material. Determination of both parameters is a primary task when scientifically reasoned approach is used to designing the adhesive joint. Therefore, the most careful attention has been paid in this monograph for estimating the aforementioned values. Test liquids used and the SFE component values are listed below (Table 2.1).

TABLE 2.1 The SFE of test liquids and its acid-base and dispersive components.

Test liquid	g_ℓ^d, mJ/m²	g_ℓ^{AB}, mJ/m²	g_ℓ, mJ/m²
Water	22.0	50.2	72.2
DMFA	32.4	4.9	37.3
Glycerol	33.9	29.8	63.7
Formamide	31.8	25.7	57.5
Aniline	41.2	2.0	43.2
Phenol	37.8	2.6	40.4
DMSO	34.9	8.7	43.6
Saturated aqueous K_2CO_3 solution	34.0	70.9	104.9
α-bromonaphthalene	44.6	0	44.6
Diiodomethane	49.8	2.4	52.2
Ethyleneglycol	29	19	48

In order for the adhesive system to meet the conditions under which the molecular contact is developed (one of necessary conditions for producing strong adhesive joints), for this system, in addition to the adhesive/substrate SFE values and thermodynamic work of adhesion, the kinetics of wetting and spreading should be studied to determine the equilibrium contact angle. When contact angles are measured, in spite of simplicity of the experiment, different problems concerned with wetting hysteresis, adhesive viscosity, substrate roughness, etc. often arise. Consider the main of them briefly.

2.2 KINETICS OF WETTING

2.2.1 *WETTING HYSTERESIS—ADVANCING AND RECEDING CONTACT ANGLES*

For estimating surface-energy characteristics of solid polymers and composite materials the contact angle of liquid with the known surface tension (i.e., test liquid) applied onto the surface of a solid (polymer, metal, composite material) is conventionally measured.

The contact angle is directly associated with the contact area of a drop of liquid. The less the contact angle the greater the contact area of the drop. The contact area of the drop defines the ability of a liquid to wet a solid. As a rule, adhesion and wetting take place sequentially: adhesion stipulates the relationship between a solid and a liquid in contact with it, which results in wetting.

There is immersion or contact wetting. Immersion wetting takes place when a solid is completely immersed in a liquid, as well as in other cases when the solid-liquid interface is formed. Contact wetting implies the presence of three phases – solid, liquid, and gaseous [23].

For each system the equilibrium contact angle under given conditions (temperature and pressure) has a single value [24]. However, the experiments show that the contact angles measured often depend on several additional factors and have different values. The dependence of contact angles on conditions under which they are formed is called wetting hysteresis which has become a certain phenomenon for several decades [25–28].

Upon contact wetting a liquid is advancing over a solid surface, gradually replacing the preceding (gaseous) phase. Therefore, the angle measured after the spreading had stopped is referred to as the advancing angle θa. Upon immersion wetting a bubble of gas pushes out a liquid, which withdraws from the previously wetted surface. Therefore, the angle measured at the starting moment of liquid withdrawal is referred to as the receding angle θr. The difference between the advancing and receding angles $\theta a - \theta r = \Delta\theta$ is called wetting hysteresis.

Contact angle hysteresis nearly always accompanies wetting. In spite of its apparent simplicity, hysteresis is hard to be learned since the difference in contact angles may be due to different causes, which often act simultaneously. First main historic causes of hysteresis are surface roughness and non-uniformity. This is explained by the fact that a three-phase contact line is affected, in addition to adhesive attraction and surface tension, by the third force similar to friction. When a liquid is advancing over a dry surface, this force may have the value different from that it may have when a liquid is withdrawing from the previously wetted surface. It is roughness and non-uniformity of a solid surface that are the main sources of the aforementioned force.

In the Young equation, the surface of a solid is assumed to be ideally smooth. However, real surfaces have rather complicated mircorelief consisting of peaks and valleys of different shape and size. Roughness affects contact angles by two reasons. The first reason is thermodynamic. Defects lead to the increased real surface compared with ideally smooth one. The ratio of these surface areas is referred to as the roughness coefficient β ($\beta > 1$) [29].

The second cause of hysteresis is kinetic. B.D. Summ explains it exemplifying by a simple element of roughness, a valley with isosceles triangled cross-sectional profile and a depth z. The valley influence is defined by its orientation against the direction of spreading. A liquid spreads along the valley freely. When perpendicularly oriented, the valley may stop spreading. Thermodynamic condition of spontaneous spreading implies that the SFE γ should decrease: $d\gamma / d\mathrm{x} < 0$, where x is the direction of spreading. This is in agreement with the decrease of the contact angle of drop: $d\theta/d\mathrm{x} < 0$, where $\theta = f(\mathrm{x})$ is so called the dynamic contact angle. For a smooth surface, this condition is true over the entire route. If any fracture, the process becomes more complicated since the increase of surface of the liquid entering the valley occurs to a greater extent than in case of the three-phase contact line moving over a smooth surface [30].

Hysteresis angles depend not only on thermodynamic parameters in the Eq. (2.4) but also on some other values. The most important of them are peak height and valley depth, roughness slope, a distance between roughness elements, and a volume of drop [30, 31]. Therefore, to estimate the degree of roughness of a real surface and how this parameter affects contact angles on surface wetted by test liquids is rather difficult task.

Examination of steel surfaces of two different grades in use, carried out by authors by means of scanning electron microscopy, showed that for Ст20 steel it is possible to use a model of parallel triangular valleys [30], about 2.6 μm in depth (Fig. 2.2a). For Ст3 steel, a miscellaneous surface relief is observed (Fig. 2.2b). According to the de Gennes low fluctuation model [31], randomly distributed roughnesses are considered as fluctuations against "zero" level, that is, ideally smooth surface.

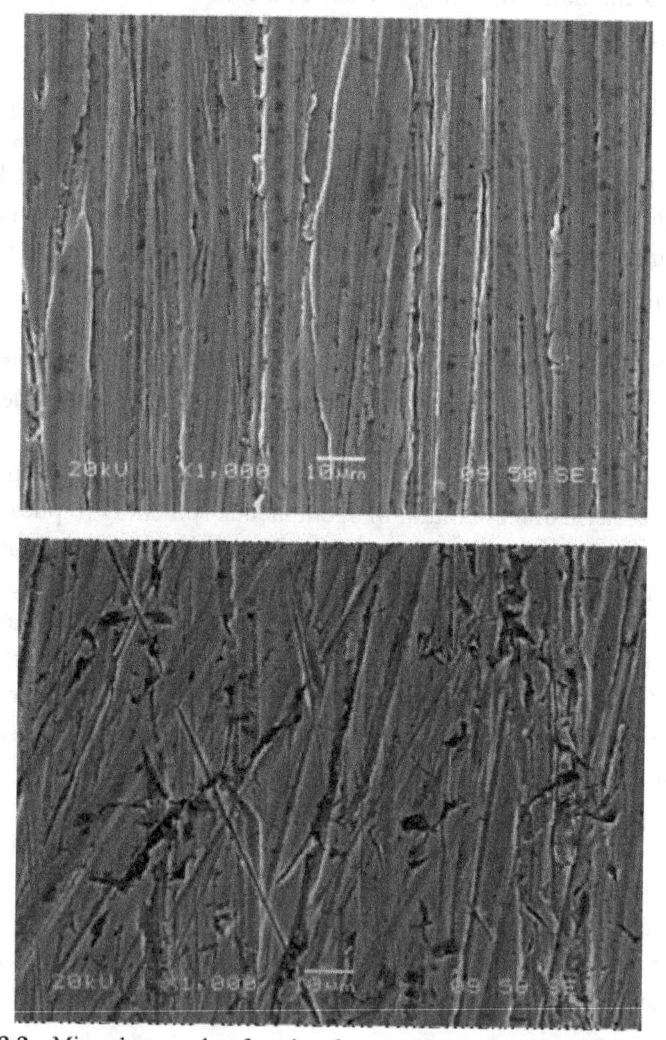

FIGURE 2.2 Microphotographs of steel surfaces: (a) St20 steel, (b) St3 steel.

This method allows the energy barrier to be calculated upon cross motion of the three-phase contact line. Significantly, such defects may obstruct the motion if their size is $z > z_{cr}$. The z_{cr} is approximately of several micrometers. Therefore, in case of smooth relief hysteresis may be caused by rather large defects only.

Defects on steel surfaces in both cases (shown in Fig. 2.2) do not exceed 5 μm that, according to the de Gennes model, is considered to be a borderline case. Therefore, provisionally speaking, defects of such metal surfaces cannot be thought to obstruct spreading of test liquid drops.

Calculation of real contact angle values by using different roughness models is complicated and, in any case, approximate. Authors exemplified the effect of the roughness degree on the cosine of contact angle for water by St3 steel.

Steel of grade St3, as the material most commonly used in adhesive joints, was subjected to special mechanical treatment for preparing samples of various purity grades. Profilograms of surfaces obtained, recorded with a 201-type profilograph, are shown in Fig. 2.3 (a, b, c). Estimation of root-mean-square deviation of peaks against the base line was made by the formula:

$$H_{ck} = \sqrt{\frac{\sum H_i^2}{n}} \tag{2.7}$$

where H_i is the ordinate of i-th peak, n is the number of peaks in the chosen part of the route. Calculations showed that purity grades of 7, 8, and 10 may be approximately assigned to the surfaces examined [32].

The roughness coefficients β were calculated for each surface as the ratio of real surface area S_{fact} to its projection onto a horizontal plane, that is, to ideal surface area (without roughnesses) S_{id}:

$$\beta = \frac{S_{\text{fact}}}{S_{\text{id}}} \tag{2.8}$$

Ultimately, for steel with various treatment degree the roughness coefficient β_1 (purity grade 7) was 1.056, β_2 (purity grade 8) was 1.027, and

β_3 (purity grade 10) was 1.01. According to this estimation, the difference between the measured cosθ and true $\cos\theta_{true}$ is 1–5%, which is within the experimental accuracy and has insignificant effect on a shape of drop.

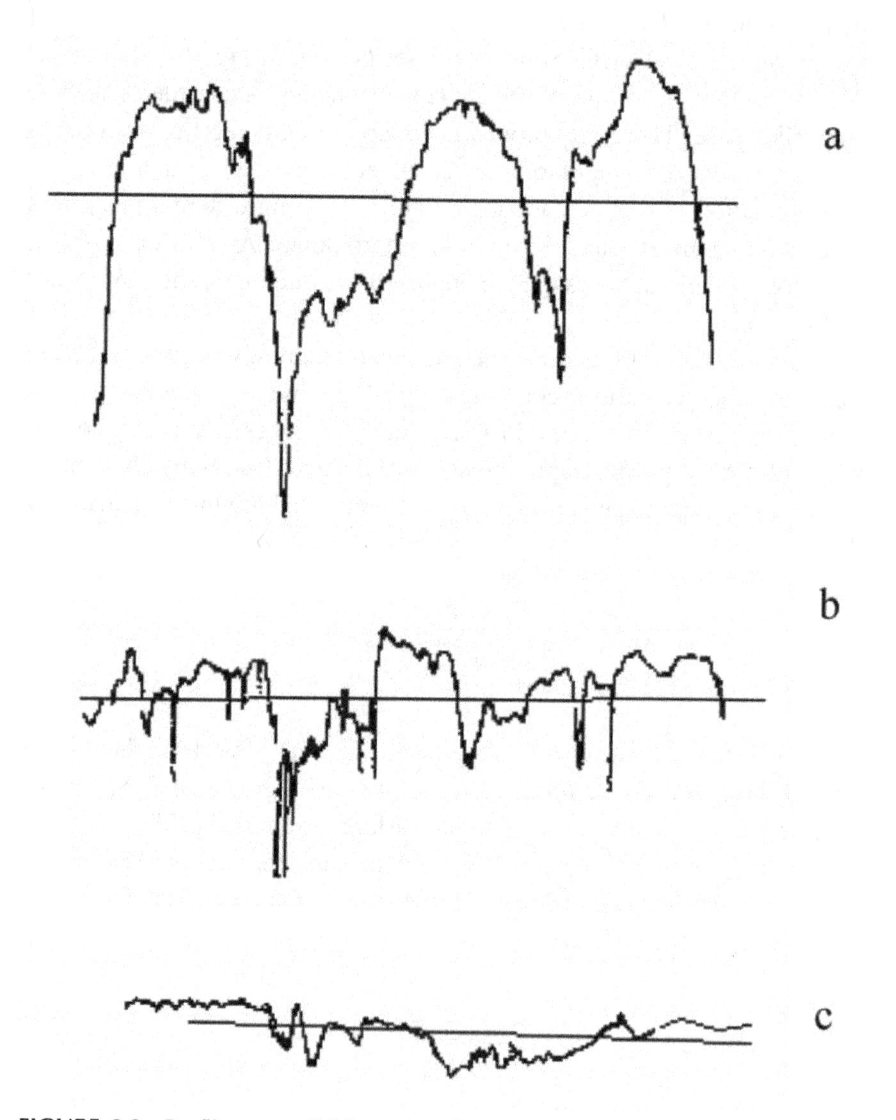

FIGURE 2.3 Profilograms of St3 steel surfaces with various roughness degrees: (a) purity grade 7, (b) purity grade 8, (c) purity grade 10.

To confirm this fact we made photographs of water drops applied to the mentioned St3 steel samples during the first three minutes. Calculation of drop parameters made thereafter evidenced that in given cases the various roughness degree does not virtually affect the contact angle values (Fig. 2.4, a–c).

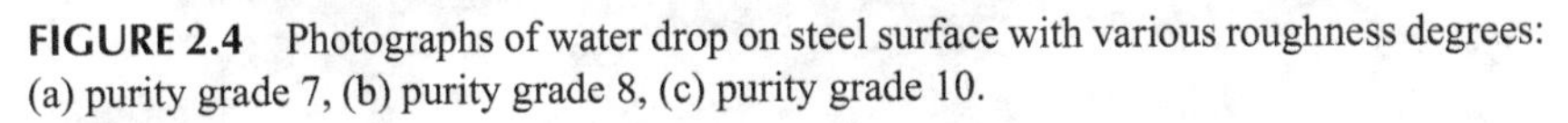

FIGURE 2.4 Photographs of water drop on steel surface with various roughness degrees: (a) purity grade 7, (b) purity grade 8, (c) purity grade 10.

As the roughness of polymer samples in all cases considered is less than that of metal samples, the roughness coefficient of organic coatings was not taken into account for further estimation of the cosines of contact angles.

The second main cause of wetting hysteresis is surface heterogeneity (non-uniformity). Surface heterogeneity may be due to two causes, namely differences in crystal structure and chemical structure. The solid heterogeneous (non-uniform) surfaces consist of regions with varied surface energy.

Structural non-uniformities are typical for polycrystalline materials which have a surface layer with a great number of little grains (monocrystals) randomly oriented to each other. The outer surface looks like a chaotic mosaic of different crystallographic faces with various surface energy values. Chemical surface non-uniformity is also typical for composites and multiphase alloys. For calculating the energy barrier height as well as the advancing and receding contact angles on non-uniform surfaces, different kinds of regular heterogeneity are used: two types of parallel strips, concentric strips, randomly arranged squares, etc. The heterogeneous surface wetting theory was developed in detail only for binary support composed of two types of regularly alternating regions [30]. Composite materials under our study, based on polyolefins, polyepoxides, and particularly rubbers, are derived from several ingredients. Therefore, physical-chemical calculation of such surfaces, taking into account inhomogeneity in composition, is still impossible. We exemplified this by microphotographs of chlorobutyl rubber (CBR) based on five ingredients, virgin (unmodified) polycarbonate, and polyepoxide formulated from ED-20+PEPA, obtained with Multi-Mode V scanning probe microscope (Veeco) (Fig. 2.5 a–c).

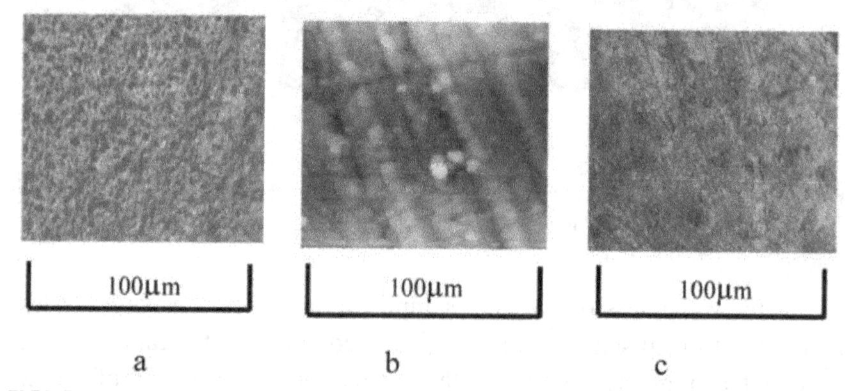

FIGURE 2.5 Microphotographs of surfaces: (a) modified rubber, (b) polycarbonate, (c) polyepoxide.

According to the estimation performed, chemical non-uniformities at composite surfaces are virtually not determined, and defects (spherical microcavities) do not exceed several micrometers, that is, the z_{cr}, in size.

The causes of hysteresis involve not only surface defects and heterogeneities. Theoretical simulation of contact angles at ideally smooth surfaces also reveals hysteresis [31, 33]. In many systems, especially at high temperatures, wetting is affected by different processes and reactions occurring when liquids interact with solids. In such systems the surface energy may vary depending on advancing or withdrawal of liquid. Of course, it results in contact angle hysteresis, which is usually referred to as physical-chemical hysteresis. Some causes of this kind of hysteresis are as follows [30].

(1) In the Young equation the vertical component of the surface tension of liquid was not taken into account. For materials with a high elastic modulus E (metals, minerals, glasses), this simplification is true because of very low deformations of materials due to this force. However, for soft materials with the modulus $E < 107$ N/m^2 (biological objects, colloidal systems, that is, gels, rubbers) such deformations are high enough. In this case, a micropeak is formed along the three-phase contact line after it has stopped to move. A peak as a typical roughness element changes the conditions of contact angle formation upon advancing and withdrawal of liquid, which results in hysteresis.

(2) The autophobic effect, that is, decreasing of the wetted surface. This phenomenon appears when an autophobic liquid is applied to the surface under study. Molecules of autophobic liquids usually have diphilic structure, that is, molecule ends are terminated by any polar group (such as OH-group) or non-polar group (alkyl group). Polyatomic alcohols are typical autophobic liquids. In case of conventional test liquids, the aforesaid refers to ethyleneglycol and glycerol.

(3) When in contact with the wetting liquid, macromolecules of polymer surface layer can gradually change their spatial structure, that is, conformation (provided that chain is flexible enough and separate segments are movable). Upon contact with water, this process leads to gradual movement of polar groups and segments to the

interface. This results in the decreased interfacial energy $\gamma_{s\ell}$ and corresponding variation the receding angle.

It was understood from Ref. [35] that contact angle hysteresis is also dependent on polymer chain configuration. The greater hysteresis was determined for amorphous polymers, and the least one – for polymers with oriented molecular chains. This may be explained by surface sorption of liquid and its penetration into polymer. Among causes of hysteresis, the molecular topography, difference between supramolecular structures on surface and in bulk of polymer, penetration of liquid into polymer and its swelling, liquid sorption and retention, as well as strong interactions between molecules of a liquid and a solid are mentioned by several authors [36–41].

Thus, there are many causes affecting the shape of a drop of liquid on a solid surface. Therefore, regarding a real three phase system, a reasonable question arises when the equilibrium contact angle is achieved and whether this achievement is possible.

2.2.2 DETERMINATION OF THE EQUILIBRIUM CONTACT ANGLE

Kamusewitz [42] as well as Della Volpe et al. [43] have recently proposed novel experimental techniques for determining the contact angle which corresponds to a "stable" or "ideal" equilibrium by developing the earlier method [44–46] based on system oscillation. The dependence of relaxation force on the depth of Teflon sample immersion in test liquid was obtained as two parallel lines, namely, the upper one is for receding angles, and the lower one is for advancing angles. The third parallel line drawn between them allowed obtaining the equilibrium angle.

Kamusewitz et al. [42] proposed to determine the equilibrium angle from a linear experimental dependence of θa and θr on hysteresis value $\Delta\theta = \theta a - \theta r$. However, even detailed studies carried out for paraffin, Teflon, and other low-energy surfaces did not result in a unique answer to a question whether the angles under measurement are truly equilibrium, or the equilibrium is metastable. There is a common opinion in up-to-date literature that all data being reported refer to a metastable state. For this case, the term "apparent" SFE (i.e., determined with "metastable" contact angles) is used. Usually, the static (unchangeable for a certain amount of time) contact angles of a sessile drop, provisionally considered as equilibrium, are measured. In other words, all the values reported in literature refer to "apparent", and if two "apparent" contact angles (i.e., advancing and receding) are more feasible to characterize a surface, why could not they be used for determining the SFE, compared with "unfeasible" Young's contact angle?

Mostly, the advancing contact angles are used for estimating the SFE. For the sake of simplicity, no liquid film is assumed to be around a drop on a solid surface. Today, the relationships between the advancing, receding, and equilibrium angles based on the Wenzel [29] or Cassie [47] equations are known, however the problem of determining the equilibrium contact angle is still unsolved.

It was found that the advancing angles remain unchanged during a certain amount of time, while the receding angles depend on time of contact with support, that is, they gradually become smaller as the contact time increases. As a possible explanation of this phenomenon the assumption was made that a liquid is absorbed and penetrates into polymer. Penetration of liquid into polymer, its retention on surface, and other similar processes result in polymer surface modification, with the SFE being different from the true value. Therefore, the receding angles represent the surface with the changed energy, and it is undesirable that the receding angles, or any function comprising both angles, be used for characterizing the SFE of surface. Many authors, for example [35], recommend using the advancing angles only.

For obtaining correct SFE values and acid-base surface characteristics by using contact angles, we plotted kinetic curves for the cosines of contact angles. Digital photographs of drops of several liquids made during 10–12 minutes starting from the moment they had been applied are shown in Figs. 2.6–2.8.

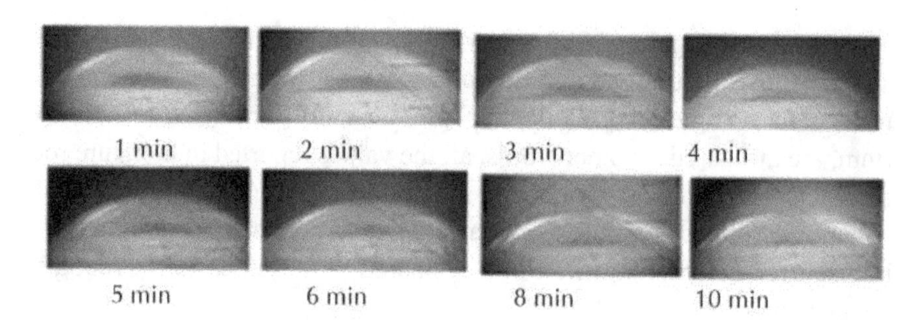

FIGURE 2.6 Aniline drops on polypropylene surface.

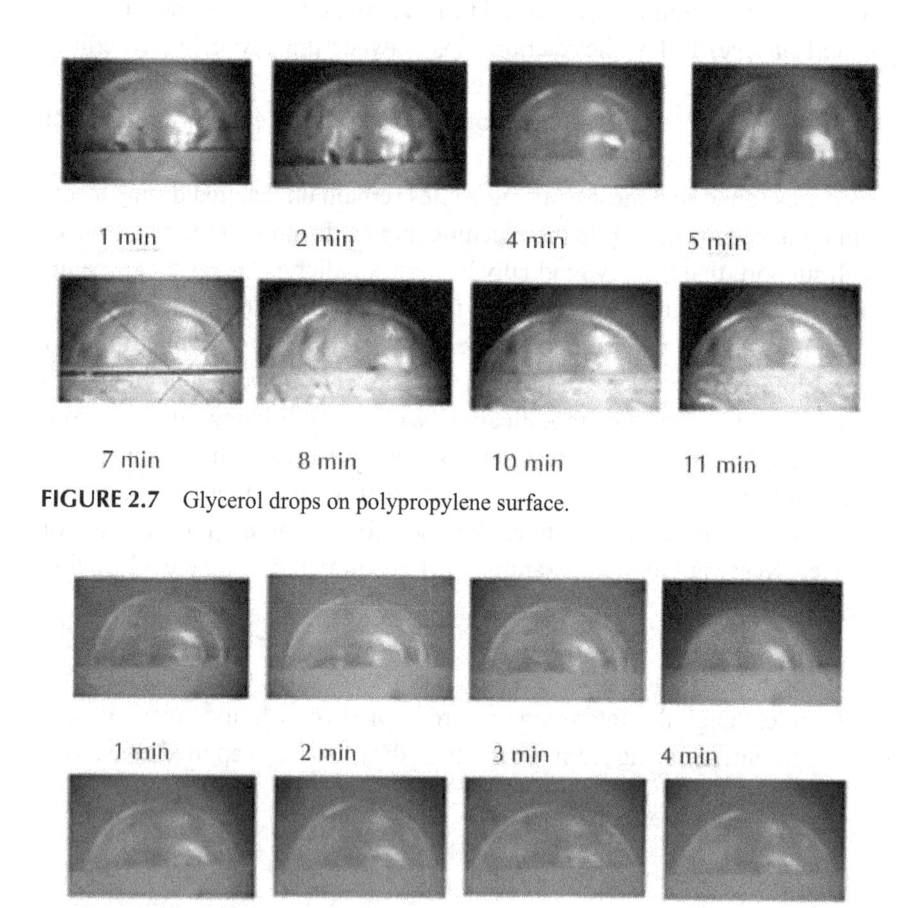

FIGURE 2.7 Glycerol drops on polypropylene surface.

FIGURE 2.8 Water drop on polypropylene surface.

Dependencies of the cosine of contact angle on time for test liquids are shown in Figs. 2.9–2.11. In all cases the same polypropylene support (the roughness coefficient is near 1) was used, so the influence of surface roughness and heterogeneity was the same in all experiments. Contact angles were measured from both sides of a drop, and the results obtained were averaged.

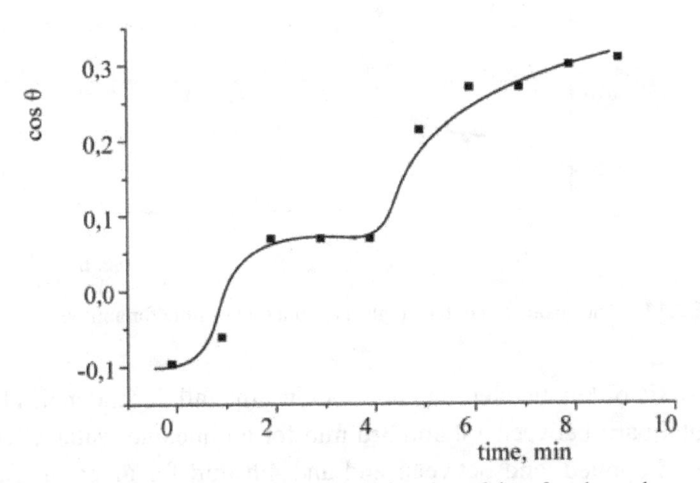

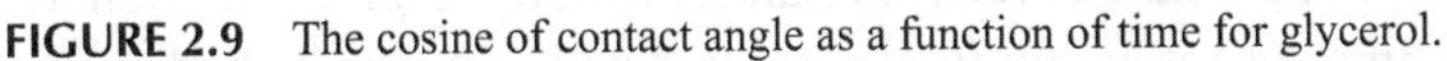

FIGURE 2.9 The cosine of contact angle as a function of time for glycerol.

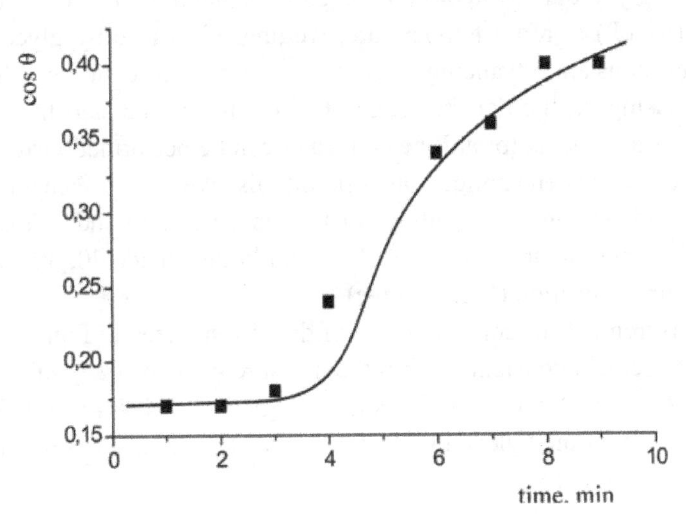

FIGURE 2.10 The cosine of contact angle as a function of time for water.

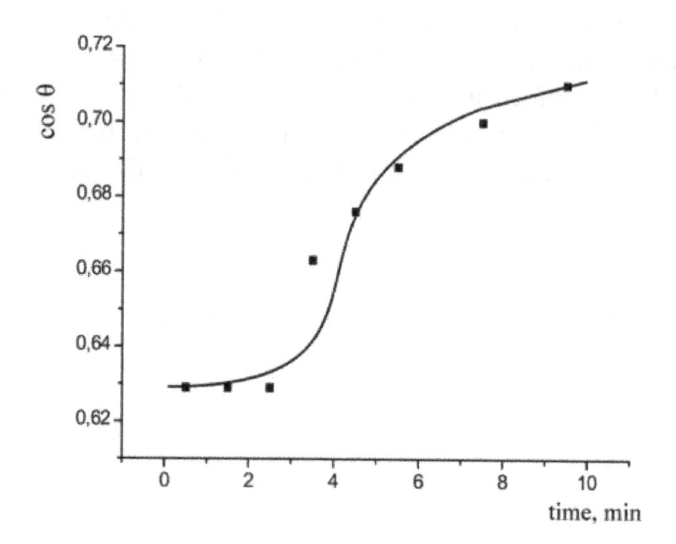

FIGURE 2.11 The cosine of contact angle as a function of time for aniline.

Analysis of kinetic wetting curves shows that all dependencies have a horizontal part: between 1st and 3rd min for aniline and water after they have been applied, and between 2nd and 4th min for glycerol. Contact angles remain constant within time periods stated. Because of higher viscosity of glycerol (~1400 cP) compared with that of aniline (3.77 cP) and water (~1 cP) at room temperature, wetting of surface by glycerol and reaching constant advancing angle values occur more slowly (during 2 min). Basing on the dependencies obtained it may be said that contact angle measurements for aniline and water can be performed virtually just after they have been applied. Other liquids also wet the surface under test, with the advancing angle getting constant in several seconds. This is evidenced by photographs of drops of several liquids made 20, 40, and 60 s after their application (Fig. 2.12a–c).

Measurements of contact angles of drops visualized in Fig. 2.12 show that they remain constant (within the measurement accuracy of ±1°) during the time mentioned. So, the contact angle is 46°-47°-47°, 44°50,' and 23°-24°-23° for methylene iodide, formamide (in all three cases), and phenol correspondingly.

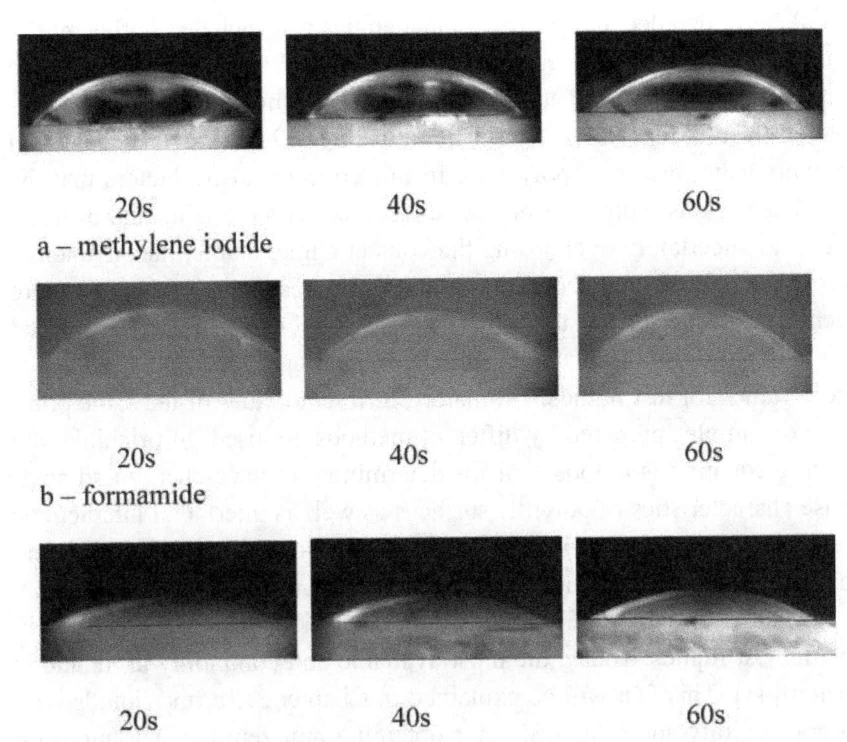

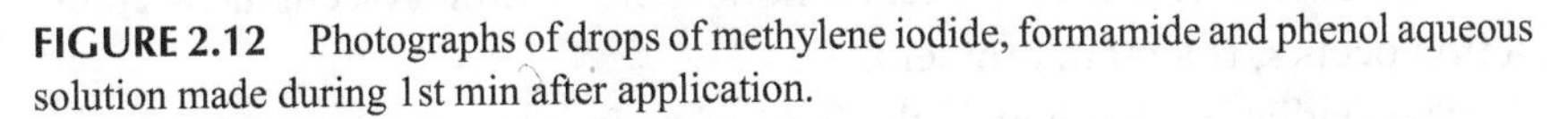

FIGURE 2.12 Photographs of drops of methylene iodide, formamide and phenol aqueous solution made during 1st min after application.

The method used here for estimating the cosine of contact angle by measuring drop dimensions, such as height and diameter, finds a wide application [35, 36]. It may be used provided that the deformation of drop under gravity due to its small size can be neglected. According to the estimation performed by authors (for drops with maximal size), the diameter of drops of test liquids applied onto the test surface did not exceed 10^{-3} m, so the volume of such drops was about 8.5×10^{-10} m^3, and the gravity acting on the drop was about 10^{-9} N.

Thus, for further estimation of the cosines of contact angles the measurements of drop sizes were carried out as follows:

- Between 2nd to 4th min after application of glycerol;
- Within 1st min after application of aniline and other liquids

A fairly detailed analysis of contact angles was included in this monograph because of recent tendency in world adhesion practice to doubt the data of wetting experiments [37]. This, in turn, is motivated by discrepancies in results reported in world literature by different, even recognized authors using the same polymers. In this concern, many factors may be cited as reasons. Some reasons are concerned with experimental difficulties and uncertainty in choosing the contact angle, that is, the advancing, receding, or metastable equilibrium one. Other reasons are related to more general aspects, such as the effect of mathematic form of the theory used on the final result or the choice and determination of correct SFE component values for test liquids. Ultimately, different grades of the same polymer or samples prepared by different methods are used. In principle, the Young equation is a good tool for determining surface-energy and acid-base characteristics of polymer surfaces as well as interfacial interactions and consequent surface modification and tailoring. However, its practical application shows that such problems are far from being solved.

Another drawback of obtaining data is a lack of monopolar especially acidic test liquids. Today, the most available and commonly used acid is water [43]. This fact will be explained in Chapter 3. To find liquids with a good acidity and a high SFE for obtaining non-zero contact angles on polymer surfaces, provided that no noticeable swelling or chemical interaction occurs, is a main problem.

The leading researchers in the field of surface thermodynamics believe that contact angle measurements should be carried out not in a single laboratory but on a worldwide level [43]. A great deal of work should be done in order to obtain good and reproducible contact angle values as a starting point of calculating the required thermodynamic characteristics. This also refers to test data series. Only if such common work produces no acceptable results, the scientific community will have to reconsider applicability of the contact angle approach to solid surface measurements completely.

As a result of the aforesaid, thorough determination of a time period within which the contact angle remains unchanged for each test liquid, as well as knowledge of the dependence of the θ value on the roughness degree of surface is required. The Young equation obtained for characterizing the behavior of an ideal liquid on an ideal surface is true for wetting of any surface by any liquid. Hence, in case of measurements performed in view of all factors affecting the truth of the results obtained, it is not worth taking statements about interfacial thermodynamics and the contact angle theory as "the comedy of errors" [38, 39], occurred in scientific literature, into account.

2.3 WETTING THERMODYNAMICS

2.3.1 THERMODYNAMIC WORK OF ADHESION OF TEST LIQUIDS ON SURFACES UNDER STUDY

The dispersive and acid-base interactions make a main contribution to thermodynamic work of adhesion. According to up-to-date literature data [48], the work of adhesion on the border of the phases 1 and 2 are additively defined by two components, namely the van der Waals component W^d (referred to as the dispersive component in present work) and the acid-base component W^{AB}.

$$W_{a12} = W_{12}^{d} + W_{12}^{AB}. \tag{2.9}$$

TABLE 2.2 Thermodynamic work of adhesion of test liquids on polymer surfaces.

No	Sample	W_a, mJ/m^2							
		Water	Aniline	Phenol	Formamide	Glycerol	α-Bromo-naphthalene	K_2CO_3	Methylene iodide
1	Fluoroplastic	51.3	66.0	62.5	52.4	58.8	72.1	73.90	76.4
2	Polystyrene	56.0	83.8	80.0	65.9	79.9	88.4	87.4	98.5
3	EVA 14% VA groups	62.1	78.9	78.6	64.1	76.1	86.4	150.8	89.3
4	EVA 22% VA groups	65.0	80.6	80.5	63.7	74.2	88.8	147.6	92.7

TABLE 2.2 *(Continued)*

No	Sample	W_a, mJ/m²							
		Water	Aniline	Phenol	Formamide	Glycerol	α-Bromo-naphthalene	K_2CO_3	Methylene iodide
5	Polypropylene	70.25	80.15	76.7	71.0	74.5	85.6	85.4	80.8
6	EEA	79.2	78.0	80.0	73.4	84.9	86.2	134.6	81.2
7	PMMA	80.8	82.6	79.3	85.6	91.5	86.1	97.3	96.1
8	EBA	83.9	83.6	79.4	66.8	84.6	84.3	125.3	90.9
9	PE grade 153	94.2	78.5	76.4	78.5	67.0	86.6	126.8	90.2
10	PE grade 277–73	94.3	79.2	77.1	82.2	93.4	87.3	124.6	83.4
11	PET	95.5	83.9	80.4	87.2	100.15	89.0	127.6	100.0
12	EVA terpolymer	96.9	81.1	76.2	93.0	88.3	85.1	145.5	89.5
13	Polycarbonate	100.3	83.6	69.4	81.6	89.1	88.9	130.6	94.6
14	Polypropylene carbonate	101.8	80.9	74.2	86.2	90.7	88.7	163.5	102.1

The dispersive component of the work of adhesion is being successfully calculated from the geometric mean values γ_1^d and γ_2^d, that is, $W_{12}^d = 2(\gamma_1^d\gamma_2^d)^{1/2}$. Similarly, the attempt was made by many researchers to calculate the acid-base component (often referred to as polar) in the same manner, that is, by using the equation $W_{12}^{AB} = 2(\gamma_1^{AB}\gamma_2^{AB})^{1/2}$. It should be noted that acid-base contribution to the surface energy γ^{AB} is usually an inadequate measure of polarity, especially for liquid-liquid systems. For solid-liquid systems, however, it is this equation that is now in use, since interfacial tensions $\gamma_{s\ell}$ for solid-liquid systems can not be measured in a different experimental way, and the errors resulted from this equation may be neglected [49].

The values of thermodynamic work of adhesion obtained by us according to the Young-Dupré equation for test liquids on surfaces of coatings of different chemical nature (polyolefin, epoxy, rubber, and metal), used in protective coating technology, are listed in Tables 2.2–2.4 and serve as reference information.

Samples in Tables 2.2–2.4 are arranged in increasing order of the work of adhesion of water on test surface (i.e., hydrophilicity). Their comparative analysis shows that they are wetted by Lewis acids and bases as well as by other test liquids differently.

TABLE 2.3 Thermodynamic work of adhesion of test liquids on polyepoxide and steel surfaces.

No	Sample	W_a, mJ/m²							
		Water	Aniline	Phenol	Formamide	Glycerol	α - B r o m o - naphthalene	K_2CO_3	Methylene iodide
1	ED-20+ PEPA	120.2	86.1	80.2	104.9	98.2	88.6	179.8	97.8
2	ED-20+ AF-2M	116.5	85.8	80.0	100.0	103.9	82.9	177.5	98.9
3	PEF-3A+ AF-2M	94.2	78.2	79.0	88.5	84.2	84.1	123.0	98.9
4	ED-20+ DTB-2	87.5	82.4	78.0	76.2	78.0	86.1	131.9	88.1
5	PEF-3A+ DETA	86.2	84.9	77.2	90.3	92.6	87.7	136.9	97.1
6	ED-20+ DETA	85.8	84.7	79.8	86.4	88.5	86.3	120.4	90.4
7	St3 steel	72.6	70.2	74.0	87.5	64.9	85.3	159.1	85.3

TABLE 2.4 Thermodynamic work of adhesion of test liquids on synthetic rubber surfaces.

No	Sample	W_a, mJ/m²							
		Water	Aniline	Phenol	Formamide	Glycerol	α-Bromo-naphthalene	K_2CO_3	Methylene iodide
1	IIR	59.1	65.8	69.2	72.3	64.4	78.4	96.6	73.9
2	IR	65.2	74.0	70.4	48.4	59.7	79.2	123.8	89.2
3	EPDM	67.6	68.6	65.6	57.2	56.7	75.3	101.5	82.8
4	CIIR	70.1	76.7	73.5	66.2	59.8	81.7	128.2	81.3
5	BR	78.5	75.5	72.5	63.5	62.5	80.2	139.5	85.6
6	SKS-30	82.7	77.2	72.4	74.4	73.4	83.0	114.0	86.7
7	Stereoregular BR	91.6	78.5	73.4	74.5	77.8	82.9	153.1	83.5

Since test bases (aniline and formamide) simulate acid-base interaction with functional groups of surfaces of acidic nature, the unmodified polymers (Table 2.1) may be arranged in increasing order of such interaction in the following row using sample numbers: $1 < 3{,}4 < 2 < 5{,}6{,}8 < 9 < 10 < 7 < 13 < 11{,}14 < 12$. Similarly, in case of test acids (phenol and glycerol) a row $1 < 9 < 5 < 3{,}4 < 13 < 2 < 8 < 12{,}14{,}6 < 10 < 7 < 11$ may be obtained. Basing on the W_a values obtained, all samples are amphipathic to some extent by having surface functional groups of both acidic and basic nature. It is seen from Table 2.1 that sample No.11 (polyethylene terephthalate) shows the greatest functional activity, while sample No.1 (fluoroplastic) shows the least one. The latter is evident enough because of unique physical-chemical properties of fluoroplastic. It is not wetted by water and most organic solvents, show poor adhesive properties and a low SFE value.

As seen from Table 2.3, polyepoxides Nos. 1 and 2 show the greatest functional activity, while steel shows the least one. The samples may be arranged in increasing order of interaction with acids (glycerol and phenol) in the following row: 7< 4 < 3 < 6 < 5 < 1 < 2. Similarly, in case of test bases a row 7 < 3 < 4 < 6 < 5 < 6 < 2 < 1 is obtained. So, these samples are also amphipathic. A slight dissymmetry of rows is concerned with various amphipathicity of samples Nos. 1–2 and 3–4.

Samples Nos. 1, 2, 4, and 6 are epoxy primers based on ED-20 epoxy resin, cured by different curing agents at room temperature. Primers based on PEF-3A epoxy urethane rubber are also included in Table 2.3. By comparing samples based on the same curing agents (sample pairs Nos. 2–3 and 5–6), it can be seen that amino phenol curing leads to greater polarity of epoxy sample, while DETA curing leads to greater polarity of epoxy urethane sample.

Synthetic rubbers in Table 2.3 are also arranged in increasing order of hydrophilicity. According to interaction with test acids and bases, samples Nos. 6 and 7 show the greatest functional activity.

The analysis of literature data showed that wetting hysteresis value itself $\Delta\theta = \theta a - \theta r$ rarely exceeds 10° being usually of 5–7° on average [50]. Since the equilibrium contact angle values fall within this range, we introduce a systematic error of ~5° (at most), using the advancing angle to calculate thermodynamic work of adhesion. In this case, the cosines of large angles close to 90° will differ from the equilibrium ones by about 0.08 (downward), while the cosines of small angles close to 0° will differ by 0.01. The error introduced in the equation for calculating thermodynamic work of adhesion (2.6) may vary from 0.5 to 8% for the same liquid. The maximal error of 8% is very rarely observed situation of zero equilibrium angle. In this case, as stated earlier, the Young-Dupré equation cannot be used. On average, the error of such measurements is 2–5%. We make little account of such deviations of the estimated value from the true one in further comparative analysis of surfaces.

The performed analysis of the W_a values shows that this parameter does not allow making unique conclusions about the nature of surface and the tendency of its change, since the W_a depends, first, on the γ_ℓ of test liquid itself and, second, on the unknown interfacial energy $\gamma_{s\ell}$, which is hard to be determined experimentally. The data obtained does not allow the comparative analysis of the surfaces tested to reveal their acid-base interaction ability, because the W_a values for test acids in case of the same polymer are often close to those for test bases. We can do no more than establish a fact that alteration in the W_a is likely to be related to alterations in γ_s and $\gamma_{s\ell}$ according to the equation 2.3. Thus, the increase in the W_a may be caused by either increased γ_s, or decreased $\gamma_{s\ell}$, or by combination of these. The last one is just observed, since the surfaces under study, especially rubbers, are mostly low-energy, and $\gamma_s < \gamma_\ell$ as a rule. The task of enhancing adhesive interaction generally implies the increase of polymer-based adhesive functionality, that is, the increase in the γ_s. When the SFE values of adhesive and adhered are coming close, their interfacial energy is known to decrease [3].

Thus, for explaining the observed wetting of samples by test liquids it is necessary that their SFE, the thermodynamic work of adhesion is related to, be determined. The SFE is the most important characteristic of both single surface and adhesive joint in whole, being directly responsible for a potential interfacial interaction ability of material. The adhesion interaction ability of materials can be inferred by the known values of SFE components. The SFE depends on chemical nature and some properties of test material itself, being therefore more informative than the work of adhesion.

2.3.2 SURFACE FREE ENERGY AND ITS COMPONENTS

The detailed examination of forces affecting interfacial interaction was undertaken by Fowkes [19, 51–52], who proposed, in view of the adsorptive adhesion theory, the division of the SFE into components corresponding to main types of interaction and formulated the additivity concept for SFE components.

Fowkes assumed that the SFE of a solid generally consists of the following components:

$$\gamma = \gamma^d + + \gamma^i + \gamma^p + \gamma^h + \gamma^\pi + \gamma^{ad} + \gamma^e. \tag{2.10}$$

where d stands for the dispersive component, i – the inductive dipole-dipole component, p – the dipole-dipole component, h – the hydrogen component, π – π-links or the interaction by forming π-complexes (i.e., complexes or substances in which bonding between electron acceptor (cation, metal, metal salt) and electron donor (unsaturated or aromatic compound) occurs with participation of π-electrons of donor group), *e* – the electrostatic component, *ad* – the donor-acceptor interaction.

According to van Oss et al. [53, 54], the first three members in the equation 2.10 constitute the Lifshitz–van der Waals component:

$$\gamma^{LW} = \gamma^{d} + \gamma^{i} + \gamma^{p}. \quad (2.11)$$

Here, the energies of the inductive and polar interactions γ^{i} and γ^{p} are the second smallness order values compared with the dispersive and hydrogen interactions. Therefore, only the γ^{d} value will be then used, but the γ^{i} and γ^{p} values will be neglected. Fowkes and Mostafa combined the last four members in the equation 2.10 γ^{h}, γ^{π}, γ^{ad}, and γ^{e} into the acid-base component [17]:

$$\gamma^{AB} = \gamma^{h} + \gamma^{\pi} + \gamma^{ad} + \gamma^{e}. \quad (2.12)$$

So, in most cases, the SFE is additively defined by two constituents, namely the dispersive and acid-base components:

$$\gamma = \gamma^{d} + \gamma^{AB}. \quad (2.13)$$

Mathematic form of the first component was developed within the framework of the Good-Girifalco theory [55] resulted from revelation of formal similarity between the Hildebrand theory about the energy of mixing nonelectrolyte solutions and the free interfacial energy theory. Using the "regular" solution concept [56] together with the London dispersion forces theory [57] resulted in the geometric-mean theory [58] as applied to the free adhesion energy. If the forces of molecular interaction of phases 1 and 2 are of the dispersive nature only, the Lifshitz–van der Waals component of each pure phase may be expressed as [58, 59]:

$$\gamma_1^{LW} = \frac{3\alpha_1^2 I_1}{4r_{01}^2} \quad \gamma_2^{LW} = \frac{3\alpha_2^2 I_2}{4r_{02}^2} \quad (2.14)$$

where: α is the polarizability, I is the ionization potential, r_0 is the equilibrium intermolecular distance. For the two immiscible substances 1 and 2:

$$\gamma_{12}^{LW} = \frac{3\alpha_1\alpha_2 I_1 I_2}{4\left(I_1 + I_2\right) r_{012}^2} \tag{2.15}$$

For the free adhesion and cohesion energies Good applied the Berthelot geometric mean rule [60] and derived the expression:

$$\frac{-\Delta G_{12}^{A}}{\left(\Delta G_1^K \Delta G_2^K\right)^{1/2}} = f \tag{2.16}$$

where: $\Delta G^A_{12} = \gamma_{12} - \gamma_1 - \gamma_2$ is the free energy of adhesive interaction of phases 1 and 2; $\Delta G^K_{1,2} = 2\gamma_{1,2}$ is the free cohesion energy of any pure phase. Considerations were reasoned for the dispersive (Lifshitz–van der Waals) SFE component. By inserting expressions for the free adhesion and cohesion energies in the Eq. (2.16), taking the factor $f \approx 1$ for polymeric substances into account, the following was obtained:

$$\gamma_{12}^d = \gamma_1^d + \gamma_2^d - 2\sqrt{\gamma_1^d \gamma_2^d} \tag{2.17}$$

$$= \left(\sqrt{\gamma_1^d} - \sqrt{\gamma_2^d}\right)^2$$

The dispersive interaction is most versatile, since it is stipulated by interaction of molecules between themselves due to their momentary microdipoles. When molecules are approaching to each other, the orientation of microdipoles becomes no longer independent, but their appearance or disappearance in different molecules occurs in time with each other, which results in interaction of momentary dipoles. More precisely, timed appearance or disappearance of microdipoles in different molecules results in their attraction. The energy of interaction of non-polar molecules is the averaged result of interaction of such momentary dipoles.

Further, Fowkes proposed [49] that for a solid-liquid system, the geometric mean value of the dispersive component of the work of adhesion $W_a^d = 2(\gamma_s^d \gamma_\ell^d)^{1/2}$ allows the interfacial interaction caused only be dispersive forces to be truly predicted, and, neglecting the pressure of spreading of vapor adsorbed onto a solid surface, derived:

$$\cos\ \theta = \frac{-\gamma_l + 2\left(\gamma_s^d \times \gamma_l^d\right)^{1/2}}{\gamma_l} \quad (2.18)$$

In fact, for the dispersive SFE components, the geometric mean expression for energies of two phases implies equal involvement of those phases into interaction. This means that all potentially existing at the moment momentary dipoles of the one phase participate in electrostatic interaction with momentary dipoles of another phase only. This is true due to a versatile character of dispersive forces, provided that two phases are in a full contact. This Fowkes method is used for calculating the dispersive SFE component values γ_s of solid non-polar surfaces which meet the following requirement $\gamma_s^{AB} = 0$ [19, 49].

2.3.3 CALCULATION OF THE DISPERSIVE SFE COMPONENT USING NEUTRAL TEST LIQUIDS

We shall consider the reliability of results being obtained due to estimation of the dispersive SFE components with neutral test liquids. For a neutral test liquid (squalene, methylene iodide, α-bromonaphthalene), when $\gamma_\ell = \gamma_\ell^d$ и $W^{AB} = 0$, the expression $W_a^d = W_a = \gamma_\ell^d(1+\cos\theta)$ is true. Among all test liquids used by us (Table 2.1), α-bromonaphthalene and methylene iodide refer to predominantly neutral and have the γ_ℓ^{AB} of 2.4 and 0 mJ/m^2 correspondingly. In several literature sources [53], methylene iodide is considered to be completely dispersive, with the acid-base component being equal to zero. Some authors assign a slight acidity to it. Thermodynamic work of adhesion of given liquids on test surfaces will include almost only the dispersive component which yields the dispersive SFE component of given surfaces due to calculations according to the Eq. (2.9):

$$\gamma_\ell^d(1+\cos\theta) = 2(\gamma_s^d\,\gamma_\ell^d)^{1/2},\ \gamma_s^d = \gamma_\ell^d((1+\cos\,\theta)/2)^2. \quad (2.19)$$

The γ_s^d values of different types of some surfaces under study, such as unmodified polymers, epoxy primers, metals, calculated by us using methylene iodide and α-bromonaphthalene as neutral liquids, are systematized in Table 2.5. Similar calculations carried out for other samples yield in all cases the greater dispersive SFE component values compared with those reported in literature.

TABLE 2.5 The dispersive SFE components of different surfaces calculated by using neutral test liquids.

Sample	γ_s^d			γ_s^d from literature sources [3, 50]
	Obtained with α-bromo naphthalene	Obtained with methylene iodide	Mean value	
HDPE-153	37.3	41.7	39.5	31.1
SEVA-22	38.6	36.4	37.5	30.2
ED-20+ PEF-3A+ PEPA	44.25	45.6	44.9	37.2
PEF-3A+ PEPA	41.0	44.9	43.0	34.3
ED-20+PEPA	39.4	32.3	35.8	41.2
PMMA	44.0	38.3	41.15	41.0
CIIR	37.4	30.2	33.8	—
PTFE	29.6	22.7	26.15	20.7
Polycarbonate	44.3	40.85	42.6	—
Polypropylene	41.1	29.9	35.5	29.4
Polystyrene	43.8	44.3	44.05	41.9
PET	44.15	45.9	45.0	41.8
St3 steel	40.7	33.1	36.8	35.4

Possibly, these discrepancies may be explained by the fact that simulation of dispersive interactions on surface yields reliable results only in case of fully neutral surfaces (such as fluoroplastic or thermally unoxidized polyethylene). In our case, the samples (even HDPE-153) show slight-to-noticeable functional activity, and the expression for W_a is likely to include also the acid-base member W^{AB}. At the same time, almost every liquid shows self-association ability, that is, some amphoterism. In particular, as mentioned earlier, methylene iodide is characterized by a slight (2.4 mJ/m^2) acid-base component. If a surface has inherent (even if a slight) functional activity, even the very negligible ability of liquid to form acid-base bonds on test surface will be shown. This would result in the increased cosines of contact angles and, hence, in the increased γ_s^d. Finally, the very probable cause of discrepancies is most likely to be the differing nature of sample surfaces. The same domestic and imported polymer grades are known to differ in properties. Also, there are most diverse grades of almost all substances stated produced in Russia as well as abroad. An extra cause may be different preparation of surfaces. Nevertheless, some values obtained by us are close enough to those reported in literature, which also cannot be considered as absolutely accurate.

Thus, the aforementioned method is poorly suitable for solving the problem of the dispersive SFE component determination, so much more that the most important role in establishing the adhesive contact is played by the acid-base SFE component besides the dispersive one. In this concern, we consider other approaches to their determination using wetting energy and a great number of test liquids of different nature, that is, neutral, amphoteric, hard and soft Lewis acids and bases to be used for simulating all potential surface interactions.

2.3.4 THEORETICAL DETERMINATION OF THE ACID-BASE SFE COMPONENT

First, the difference between the γ^p and γ^{AB} should be stressed. Some authors do not differentiate these items, which often leads to errors in data. Mathematic form of the energy of dipole-dipole or polar intermolecular interaction γ^p for dipoles free to rotate was given by Keesom [61]. Suppose μ is a dipole moment of molecules of the phase 1, then:

$$\gamma_1^P = 2\mu_1 / 3kTr_{01}^2 \quad (2.20)$$

Similar expression may be written for the phase 2. The energy of interfacial dipole-dipole interaction may be expressed as:

$$\gamma_{12}^{P} = 2\mu_1^2\mu_2^2 / 3kTr_{012}^2 \tag{2.21}$$

Similarly to Eqs. (2.16)–(2.17), the following may be yielded:

$$\gamma_{12}^{P} = \gamma_1^{P} + \gamma_2^{P} - 2\sqrt{\gamma_1^{P}\gamma_2^{P}} \tag{2.22}$$

$$= \left(\sqrt{\gamma_1^{P}} - \sqrt{\gamma_2^{P}}\right)^2$$

The dispersive and dipole-dipole interactions are not fully independent on each other. As a result, the component responsible for inductive interaction takes place. The inductive SFE component of each phase may be written as [57]:

$$\gamma_1^{i} = \frac{\alpha_1\mu_1^2}{r_0^6} \qquad \gamma_2^{i} = \frac{\alpha_2\mu_2^2}{r_0^6}; \tag{2.23}$$

Then, the interfacial inductive component is as follows [57, 62]:

$$\gamma_{12}^{i} = \frac{\left(\alpha_1\mu_1^2 + \alpha_2\mu_2^2\right)}{2r_{012}^2} \tag{2.24}$$

Fowkes estimated relative contributions of the dispersive, dipole-dipole, inductive, and acid-base interactions. He made a conclusion that the γ^{p} and γ^{i} are the second smallness order values compared with the dispersive component, and the acid-base member is comparable to the dispersive one by value.

Thus, the polar SFE component is stipulated by interaction of constant and induced dipoles, whereas the acid-base component is caused by donor-acceptor interaction.

As mentioned above, calculation of the dispersive SFE component may be successfully performed by using the W_a^{d} and geometric mean product of dispersive SFE components of phases in contact. Since the obtained results are reliable, many researchers made an attempt to calculate the acid-base SFE component in a like manner. Owens and Wendt [63] as well as Kaelble and Uy [64] proposed the acid-base component of the work of adhesion to be written as the geometric mean $2(\gamma_s^{AB} \times \gamma_\ell^{AB})^{1/2}$ and derived the formula which may be used for calculating the SFE in solid-liquid system:

$$W_a = \gamma_l \left(1 + \cos\ \theta\right) = 2\left(\gamma_s^d \times \gamma_l^d\right)^{\frac{1}{2}} + 2\left(\gamma_s^{AB} \times \gamma_l^{AB}\right)^{\frac{1}{2}} \tag{2.25}$$

Results obtained by using the Eq. (2.25) are in a good agreement with similar values derived by other methods. However, some authors [52, 65, 66] made critical notes in regard to using the geometric mean rule for determining the acid-base component because of internal forces in liquids and solids. Fowkes considered using this expression for liquid-liquid system as inadmissible, since interfacial tension values thus obtained are far from real. In fact, almost every substance, particularly liquids, is prone to self-association to some extent due to its amphoterism. There are few fully neutral liquid substances. Generally, these are saturated hydrocarbons and some organic solvents. The acid-base sites present in liquid are partly used for the formation of "internal" cohesive acid-base bonds. Therefore, the geometric mean rule may be used for estimating acid-base SFE components with assumptions only. Many researchers apply this equation to determine the γ_s^{AB} and W^{AB} values for solid-liquid system mostly because interfacial tension values $\gamma_{s\ell}$ cannot be determined experimentally, the simplicity of practical application is evident, and errors resulting from such approximate calculations are thought to be insignificant. Thus, provided that acid properties of test Lewis bases and base properties of test acids are neglected, the Owens-Wendt equation for calculating the γ_s^{AB} by measuring contact angles on test surface wetted by these liquids may be applied. The error thus introduced would lower the obtained γ_s^{AB} values depending on the amphoterism of liquids used.

There is an opinion that other expressions, such as the harmonic or harmonic-geometric mean equations (2.26) and (2.27), would be preferred for estimating the dispersive and polar interactions [67]:

$$\gamma_\ell(1+\cos\theta)/4 = \gamma_s^d \gamma_\ell^d/(\gamma_s^d + \gamma_\ell^d) + \gamma_s^{AB} \gamma_\ell^{AB}/(\gamma_s^{AB} + \gamma_\ell^{AB}), \tag{2.26}$$

$$\gamma_\ell(1+\cos\theta)/2 = (\gamma_s^d \gamma_\ell^d)^{1/2} + 2\,\gamma_s^{AB} \gamma_\ell^{AB}/(\gamma_s^{AB} + \gamma_\ell^{AB}). \tag{2.27}$$

However, it is the geometric mean the appropriateness of using which for estimating both the dispersive and non-dispersive forces was verified theoretically and experimentally.

Relatively new, "revolutionary" solution of this problem was proposed in the end of 1980s by van Oss et al. [68–70] who assumed that the acidic groups of one phase interact with the basic ones of another phase in case of interfacial acid-base interaction. If a substance shows both acidic and basic properties, the so-called cross coupling occurs. Some substances, such as ketones, act as bases when participating in the hydrogen bond formation. These are referred to as mono polar, or basic substances. On the contrary, substances like chloroform are acidic. Many substances, such as water, are bipolar (amphipathic). If phases are neutral, and both phases have only basic (or acidic) groups, the acid-base interactions do not occur. Van Oss suggested saving the geometric mean in his approach, but the geometric mean should be read as contribution of new values to the SFE. For a pure bipolar substance n, we have:

$$\gamma_n^{AB} = 2\sqrt{\gamma_n^+ \gamma_n^-} \tag{2.28}$$

where: γ^+ and γ^- are non-additive SFE components of a liquid or a solid responsible for electron-acceptor (acidic) and electron-donor proton-donor (electron-acceptor), and basic, or proton-acceptor (electron- $\gamma_n^{AB} = 0$ donor) parameters correspondingly. For a strongly monopolar substance, $\gamma^+ = 0$ or $\gamma^- = 0$, and then. In case of two phases, the following combinatorial rule for the acid-base component was proposed [53, 58, 71]:

$$\gamma_{12}^{AB} = 2(\sqrt{\gamma_1^+} - \sqrt{\gamma_2^+})(\sqrt{\gamma_1^-} - \sqrt{\gamma_2^-}) \tag{2.29}$$

In contrast to the dispersive and dipole-dipole forces, the acid-base interaction is one of those the components of which are non-additive and rather complement each other. This combinatorial rule is based on the Collman [72, 73] and Hobza-Zaharadnik [74] quantum theory of hydrogen bonds.

Taking into account the equations mentioned, the expression for interaction of two polar phases may be written as:

$$\gamma_{12} = (\sqrt{\gamma_1^{LW}} - \sqrt{\gamma_2^{LW}})^2 + 2(\sqrt{\gamma_1^+} - \sqrt{\gamma_2^+})(\sqrt{\gamma_1^-} - \sqrt{\gamma_2^-}) \qquad (2.30)$$

To return to the question of difference between the terms "polar component" and "acid-base components", it may be said that the dipole-dipole SFE component of a pure liquid is always positive, but the acid-base SFE component of a strongly monopolar liquid is equal to zero. When two polar liquids interact with each other, the acid-base SFE component, according to the Eq. (2.29), may be even negative. Therefore, equating the acid-base component with polar one is inadmissible.

From the van Oss approach it rather unexpectedly follows that provided a surface has, for example, only active functional groups of acidic nature, and $\gamma^- = 0$, then the acid-base SFE component of such material is equal to zero. We will analyze this statement. According to the surface energy definition, it is a measure of either uncompensated intermolecular forces in the surface (interfacial) layer or excessive surface energy in the surface layer compared with that in bulk of phases. All seven SFE components mentioned before the Eq. (2.10) are likely to have the same physical meaning but refer each to a specific kind of interaction. It makes sense to assume that if any component is equal to zero, then such kind of interaction in a given phase will be absent. If $\gamma^{AB} = 0$, should we conclude, basing on the aforementioned speculations, that such a surface is not able to acid-base interaction? It is not the case, since a surface may have, as said earlier, either acidic or basic sites separately. If adhesive (or test liquid) with opposite functionality is applied onto such surface, then both good interaction and wetting should occur. In other words, the fact that the γ^{AB} is equal to zero does not mean that the surface is fully neutral. It is necessary but insufficient neutrality condition. In order for surface to be considered as fully neutral, the conditions $\gamma^+ = 0$ and $\gamma^- = 0$, and hence $\gamma^{AB} = 0$ should be met simultaneously.

This not very clear situation might be solved by clarifying the existing terminology. The acid-base component γ^{AB} taken as the sum of the four members in the Eq. (2.12) implies to what extent a substance is prone to "own" intramolecular acid-base interactions, and does not carry information about function nal activity or polarity of the substance. To acquire such information, the member γ^{AB} should be expressed in terms of its constituent parts γ^{+} and γ^{-} to obtain their numeric values. Therefore, the γ^{AB} itself is insufficient for estimating the acid-base nature of test surface. According to numeric value of this component we may only say whether this interaction takes place on surface and to what extent a substance itself shows amphipathic properties.

Thus, theoretical explanations forming a ground for calculating the acid-base SFE component of solids are mostly indigested and incomplete. Each of approaches proposed has some drawbacks. Even the additivity concept for SFE components itself is still questionable and, according to some authors' opinion, in conflict with the electrodynamics. At the same time, there are serious experimental and theoretical evidences in favor of both the geometric mean rule and the van Oss combinatorial rule. The key is that knowledge of SFE components of substances is necessary for realizing and predicting adhesive effects. This particularly concerns the γ^{ab}, since acid-base interactions play the main role in establishing interfacial contact. The very important advantage the acid-base interaction theory offers is the opportunity to ascertain whether a substance is acidic, basic, or amphipathic. In this concern, to find and develop correct techniques for estimating energy and acid-base properties of surface, such as SFE components and other acidity (basicity) characteristics, is a primary task. Therefore, investigations in this field are being carried out extensively. The growing database of novel techniques for estimating acid-base properties as well as modification of the existing data require systematization to select the most scientifically reasoned method within the framework of a given task. The next chapter reviews the most significant methods, comparative analysis of them, and their applicability to polymers, polymeric composite materials, and metals.

KEYWORDS

- **Good-Girifalco theory**
- **surface free energy**
- **thermodynamic potentials**
- **thermodynamic system**
- **wetting hysteresis**

CHAPTER 3

COMPARATIVE ANALYSIS OF METHODS FOR DETERMINATION OF ACID-BASE PROPERTIES OF LIQUIDS AND SOLIDS

CONTENTS

Intermolecular acid-base interactions have a significant effect on the adhesion of polymers to other materials. This makes us to find a quantitative characteristic of acid-base properties of polymer surfaces and inorganic substrates widely used in adhesive techniques.

For solving this task, there are many methods at the moment, such as, inverse gas chromatography, microcalorimetry, ellipsometry, IR and NMR spectroscopy, etc. Recent studies allow characterization of a specific acid-base interaction basing on contact angles of solid smooth surfaces wetted by liquids.

The main problem of recent techniques is their limited applicability as well as frequent discrepancies in estimated values obtained by different authors using the same method. This is caused by different ways of mathematical treatment, different standards used in choosing surfaces, and different estimation of measurement errors. Discrepancies observed even in recognized authors' results indicate the necessity of developing laboratory tests for determination of a true nature of substances with acid and base terms in more detail. Such developments would be important especially for acidic and basic test liquids which themselves are used for determining surface behavior of other wetted materials. A correct choice of liquids to measure the contact angles implies a series of liquids simulating all possible typical interactions (dispersive, basic, and acidic).

This chapter describes all existing domestic and foreign techniques for quantitative estimation of acid-base behavior on solid surfaces and analysis thereof. The questions related to the interpretation of quantitative estimations mentioned and possibility of using them in practice are also considered, as well as the results obtained by authors using several methods listed are included.

3.1 THE DRAGO METHOD

Millikan interpreted acid-base interactions as the formation of charge-transfer complexes, with two constituents of interaction energy being electrostatic and covalent [75]. Realizing correctness of this approach in 1971, R.S. Drago derived his four-parameter equation for the enthalpy of formation of different Lewis acid-base systems [76]:

$$-\Delta H = E_A E_B + C_A C_B, \quad (3.1)$$

where both E_A and C_A refer to an acid, and both E_B and C_B refer to a base. The E_A and E_B parameters stand for electrostatic interaction ability of an acid and a base correspondingly, whereas the C_A and C_B stand for their ability to form covalent bonds. Drago calculated electrostatic and covalent contributions and estimated the nature of substances tested according to the Eq. (3.1) using the known acid-base interaction heat values as applied to 31 Lewis acids and 43 Lewis bases. He proposed a series of parameters for acids and bases which allow an accurate calculation of 280 mixing enthalpies as well as predicting 1200 interaction enthalpies. It should be noted that calorimetric heats of acid-base interaction, used for calculating the Drago constants, were measured in diluted solutions based on a neutral solvent (cyclohexane and carbon tetrachloride). The measured heats of van der Waals interactions of solute molecules with solvent molecules do not carry information directly on van der Waals interactions of an acid with a base.

Unfortunately, Drago disregarded a bipolar nature of many substances in his investigations, for example, the basic properties of hydrogen-bonded liquids (such as water and alcohols) and the acidic properties of amines, amides, ketones, and dimethylsulfoxide. Similarly, many other researchers are neglecting the acidity of their model bases and the basicity of their model acids. Reference Drago constants $C_{A'}$ $C_{B'}$ $E_{A'}$ and E_B being successfully used in world adhesion studies were determined with these assumptions. Nevertheless, the acid-base interaction degree estimated by the Drago equation differs from the experimental data by less than 5%. The possibility to predict the efficiency of interfacial interaction in adhesive systems by using the Drago method is slightly limited due to the lacking data on the C and E values for polymers and substrates.

3.2 THE GUTMANN METHOD

Gutmann et al. also investigated acidic and basic properties of organic liquids and estimated both acidity and basicity degrees for them, assuming that all organic liquids are prone, to some extent, to a specific acid-base self-association [77]. The strength of bases was characterized by calorimetric heats of acid-base coordination interaction of organic substances with antimony pentachloride in diluted 1,2-dichloroethane solutions. The reaction enthalpy reversed in sign was defined as the donor number D_N measured in kcal/mol. The acceptor numbers A_N were estimated by measuring ^{31}P NMR chemical shifts, which were observed upon dissolving triethylphosphine oxide as the test substance in corresponding solvent. For the chemical shift of triethylphosphine oxide in *n*-hexane, the A_N was taken as zero, and for the diluted antimony pentachloride solution in 1,2-dichloroethane it was taken as 100.

For calculating the A_N, ^{31}P NMR chemical shifts were determined not for solutions of test liquids but for test liquids themselves. So, these shifts may be considered to result from both acid-base and van der Waals interactions. The dispersive contribution to the acceptor number (A_N^d) was calculated from surface tension measurements followed by calculation of the difference $A_N - A_N^d = A_N^*$ which yields the heat of acid-base interaction of test liquid with triethylphosphine oxide.

It should be noted that both donor and acceptor numbers allow evaluation of either acidity or basicity value without considering both electrostatic and covalent contributions to acidity and basicity. Scales of the donor and acceptor numbers do not correlate with each other.

3.3 CALORIMETRY

Calorimetric titration is only suitable for polymers, which are well soluble in neutral solvents (such as xylene and carbon tetrachloride). For estimating van der Waals contributions to the mixing heats, both acidic and basic titrants should be as more inert as possible. Polymer solution is added drop wise to acid or base solution in the same solvent. Final partial molar mixing heats may be determined at infinite dilution. Similar measurements were successfully carried out for propylene oxide solution in benzene [78].

By means of calorimetry, quantitative thermodynamic measurements of the acidity or basicity degree can be performed, and the acid-base properties of organic and inorganic powder or film surfaces can be evaluated. Molar adsorption heats of test acids and bases interacting with surface active functional groups of such powders may be determined with high accuracy which depends on powder surface activity. If adsorption occurs from neutral solvents at moderate concentrations so that adsorptive films are saturated less than by half, the adsorption heats are the result of acid-base interactions with acidic or basic groups on surface.

Calorimetric titration of acidic or basic surface groups of powders is often carried out for powder suspensions in differential calorimeters as well as in flow microcalorimeters [79]. This titration is possible, if particles flocculate in stirred calorimetric vessel not too rapidly and have surface area on the order of 20 m^2/g which assures sufficient sensitivity. In such a way, the properties of silica and rutile suspension in cyclohexane were evaluated [80].

Flow microcalorimeters used for titration of powders are equipped with detector of slow flow concentration (UV adsorption detector) and laboratory computer for recording twice a second the adsorption heat detected by thermistor sensors [79]. Sensitivity of detecting the adsorption heat is on microcalorimetric level, so this device may be used for powders and films having less surface area, for example, glass fibers 14 μm in diameter [81] and clay powder with surface area of 1–4 m^2/g [82]. Phenol and *tert*-butylphenol are used as acidic test liquids, whereas pyridine and m-butylpyridine were used as basic ones. The molar adsorption heat value ΔH^{ads} is determined by this method, and the Drago constants E_A and C_A are then derived from graphically displayed four-parameter Drago equation, for example:

$$E_A = -\Delta H^{ads}/E_B - C_A(C_B/E_B), \tag{3.2}$$

where the Y-intercept (E_A) at C_A=0 is the $-\Delta H^{ads}/E_B$, and the slope denotes the C_B/E_B ratio. Thus, when test acids with the known E_A and C_A values are used, the C_B and E_B values of the test substance may be found. Strong and weak bases showing a noticeable difference in the C_B/E_B ratio are characterized by noticeably differing slopes to the X-axis.

Flow microcalorimetry is successfully used also for determining acidic surface groups of magnetic iron oxides, which cannot be titrated in suspension because of a strong magnetic interaction between particles which leads to a rapid flocculation.

3.4 SPECTROSCOPY

If NMR chemical shifts allow determination of the Drago constants C and E or the Gutmann donor and acceptor numbers, then IR spectral carbonyl shifts ($\Delta v_{c=o}$) can be used for estimating the acid-base interaction of:

- Ester groups of polyesters with polyacrylates;
- Carbonate groups of polycarbonates;
- Amide groups of polyamides as well as amide, ester, carbonate, and Aeto groups of test solvents [1].

The earliest studies on this point carried out with ethyl acetate showed that the frequency of carbonyl vibrations for ethyl acetate vapors shifts from 1764 cm^{-1} due to intermolecular van der Waals interactions and acid-base interactions between solvent molecules. Van der Waals contribution Δv^d is proportional to van der Waals contribution to the SFE γ^d. Acid-base contribution Δv^{ab} is proportional to acid-base interaction heat ΔH^{ab} for ethyl acetate-test acid interaction:

$$v_{c=o} = 1764\ cm^{-1} - 0.714\ \gamma^d\ cm^{-1}/mJ/m^2 + 0.99\ \Delta H^{ab}\ cm^{-1}/mJ/mol.$$

Proportionality of the $v_{c=o}$ frequency to the γ^d value was found by using neutral or non-acidic liquids, while the proportionality to the ΔH^{ab} value was found by using a series of acids or acidic solutions having the known (from calorimetry) heats of acid-base interaction with ethyl acetate and virtually the same γ^d values (26±1 mJ/m^2) [83].

Basing on IR shifts caused by interactions in polymethyl methacrylate solutions, the heats of acid-base interaction of this polymer with different test acids were determined and used for calculating the Drago constants. Thus, C_B=0.96±0.07 and E_B= 0.68±0.01 were calculated. The knowledge of the C and E values for polymers is necessary, but today only few of them are known.

IR spectroscopy was used for determining relative acidity and basicity of such polymers as polyvinyl fluoride, polyvinylidene fluoride, and polyvinyl chloride [84]. In case of interaction of these acidic polymers with ethyl acetate, spectral shifts of characteristic frequencies show that their acidity decreases in a series: polyvinyl fluoride > polyvinylidene fluoride > polyvinyl butyral > polyvinyl chloride.

Photoacoustic IR spectroscopy also allows determination of heats of acid-base interaction of both acidic and basic test liquids with powders of solid polymers. First, the initial spectrum is recorded using polymer powder. Then, the powder is placed into closable desiccator containing test acid or base vapors for 4 hours (when absorbed by polymer, the vapor surrounds the entire powder surface and interacts with functional groups of polymer). The new spectrum is recorded. Spectral shifts observed allow calculating acid-base interaction heats and, hence, the Drago constants [85].

As mentioned before, NMR spectroscopy is used for determining the acceptor number. The search of test liquids to be used in this method for estimating acid-base properties of solvents, polymers, and inorganic solids is being made extensively. By using NMR chemical shifts, the Drago constants have been found for ten polymers [86].

3.5 ELLIPSOMETRY

Smooth surfaces of polymers, metals or their oxides can be characterized by adsorption isotherms obtained by ellipsometric techniques, provided that the surface is well reflecting. The surface is placed in neutral solvent, and adsorption isotherms are obtained by using solutions of test acids and bases [87, 88]. Adsorption isotherms may be analyzed to find surface concentrations of acidic or basic groups. If isotherms are obtained for several temperatures, adsorption heats of these acids and bases can be calculated, followed by calculation of the Drago constants for a given surface.

3.6 INDICATOR METHOD

Surface acidity of inorganic powders (glues, cracking-catalysts) is conventionally determined by using basic dyeing indicators with various pK_a values (in water). Due to change in color of the adsorbed dyeing indicators, authors [89] estimated the acidity of titanium dioxide powders as well as the acidity of surface oxides of titanium alloy (before and after treatment for enhancing adhesive bonds). It should be noted that estimation of the strength of an acid or a base according to this method is approximate, since a relative strength of basic indicator dye is defined by the pH-value or by color change in water rather than in non-aqueous solvents.

3.7 INVERSE GAS CHROMATOGRAPHY

Gas chromatography is used for determination of the nature and surface properties of powders. Powders are placed in columns; the molecules of test vapors are adsorbed onto their surface. By using a homologous series of saturated hydrocarbons with the known γ_ℓ^d values, the effluent volume V_N at infinite dissolution can be determined, followed by calculation of the γ_S^d of powder, basing on the rule that the $\Delta G°$ of adsorption ($RT\ln V_N$) is equal to the product of the work of adhesion $W_{S\ell}$ by the surface area of the mole adsorbed $N\times a$ (where N is the Avogadro number, a is the surface of a single molecule adsorbed). For saturated hydrocarbons, it may be taken that $W_{S\ell}=2(\gamma_s^d\,\gamma_\ell^d)^{1/2}$, since the slope of a straight line is equal to $2(\gamma_s^d)^{1/2}$ in the dependence of ($RT\ln V_N$) on $a(\gamma_\ell^d)^{1/2}$, so the dispersive component itself is easy to be calculated. This dependence stands for van der Waals (mostly dispersive) contribution to molecular adsorption energy. When acid-base interactions lead to the increase in adsorption free energy and the $aW_{S\ell}^{ab}$ member appears, the $RT\ln V_N$ values plotted as a function of $a(\gamma_\ell^d)^{1/2}$ fall above the line and allow determining the $aW_{S\ell}^{ab}$. Finally, inverse gas chromatography may be used for determining both van der Waals and acid-base contributions to interfacial interactions. This method is more commonly used for determination of the γ_s^d and acidity or basicity of polymers and inorganic surfaces [1]. Thus, inverse gas chromatography has been applied for measuring PVC acidity as well as the acidity and basicity of calcium carbonate after chemical modification by plasma treatment [90].

For studying acid-base properties of polymer and inorganic powders and films by means of inverse gas chromatography, it is recommended to choose samples showing minimal acid-base self-association. The suitable sample is chloroform, since its self-association is only 2% of the heat of vaporization. Triethylamine or ethyl ether is more preferred than, for example, frequently used *n*-butylamine.

3.8 CHARGE TRANSFER

Electron transfer is usually takes place between different materials in contact. In case of metals, electrons go to the metal with less concentration of electrons, in other words, to more noble metal, until contact potential difference obstructing further transition is established at the border. If acidic dielectric (an electron acceptor) comes in contact with less noble metal, electrons from the Fermi level of metal will tend to take places of acid "traps" of isolator. Similarly, in case of basic dielectric and more noble metal, dielectric electrons go to metal [91–94]. Measurement of the resultant Volta potential gives information about acid-base properties of dielectrics (polymers).

3.9 WETTING TECHNIQUES

Among the existing techniques for determination of surface-energy and acid-base characteristics of smooth solid surfaces, the most preferred are multiphase equilibrium techniques based on interaction of low and high molecular weight compounds. The most commonly used is the sessile drop method. Information being obtained by this method allows the properties of the surface in question to be analyzed, since, as mentioned in Chapter 2, the cosine of contact angle is related to both W_a and γ_s (2.7)–(2.9). Measurement of contact angles upon wetting by test acids and bases as well as by their solutions is a conventional research method for such hydrophobic surfaces as polymers and some minerals (graphite and talc), and for sulfides and metal oxides. Analysis of wetting energy gives equally good results for solid organic and inorganic surfaces as well as for different modifiers used for improving adhesive ability of polymers [95–96].

Several aforementioned methods used in practice for determination of acid-base surface properties by calculating the work of adhesion and the SFE are worth of consideration in detail, being the most suitable for solving problems of present work.

3.9.1 THE FOWKES-MOSTAFA METHOD

Fowkes and Mostafa [17] proposed the method, which implies using molar heats of acid-base interaction between two interacting substances $\Delta H_{s\ell}^{AB}$ and interfacial concentration of acid-base bonds n_{ab} (mol/m^2) as follows:

$$W_{S\ell}^{AB} = -f n_{ab} \Delta H_{s\ell}^{AB}, \tag{3.3}$$

where f is the constant close to 1, used for transforming interfacial acid-base interaction heat into the SFE.

This method was verified for benzene-water system with the Drago constants E_A and C_A (5.01 and 0.67 (kJ/mol)$^{1/2}$) for water and E_B and C_B (0.75 and 1.8 (kJ/mol)$^{1/2}$) for benzene. The Drago equation thus yields $\Delta H_{12}^{AB} = -5.0$ kJ/mol. If benzene molecules 0.5 nm^2 in area each are planarly situated on surface, then $n_{ab} = 3.3$ μmol/m^2 and $W_{12}^{AB} / f = 16.5$ mJ/m^2, which is very close to the value measured at 20°C:

$$W_{12}^{AB} = \gamma_1 + \gamma_2 - \gamma_{12} - 2(\gamma_1^d \gamma_2^d)^{1/2} = 72.8 + 28.9 - 35.0 - 2(22 \times 28.9)^{1/2} = 16.3 \text{ mJ/m}^2$$

Della Volpe and Siboni [97] pointed out several drawbacks of this method. Thus, Fowkes expressed the work of adhesion in terms of enthalpy of interfacial interaction, whereas the thermodynamic ratio of these two parameters under isothermal conditions takes the form $\Delta G = \Delta H - T\Delta S$. Fowkes's assumption that ΔG may be expressed in terms of enthalpy by using constant coefficient f gives rise to doubt. Furthermore, the assumption that the constant f is close to 1 corresponds to a slight variation in entropy ΔS during interface formation, which is far from being possible for all cases and cannot serve as an overall conclusion. The Gibbs free energy or interfacial interaction enthalpy can be expressed via the following ratio, which differs from the Eq. (3.3):

$$\Delta H = \Delta G + T\Delta S = \Delta G - T\frac{\partial \Delta G}{\partial T}. \tag{3.4a}$$

$$\Delta H_{adh} = W_{adh} - T\frac{\partial W_{adh}}{\partial T} = \gamma_\ell(1+\cos\theta) - T\left\{(1+\cos\theta)\frac{\partial \gamma_\ell}{\partial T} + \gamma_\ell \frac{\partial(\cos\theta)}{\partial T}\right\}. \tag{3.4b}$$

3.9.2 *WETTING BY SOLUTIONS OF TEST ACIDS AND BASES*

This method is applicable for hydrophobic surfaces and solutions of test acids and bases in a neutral liquid with high surface energy. For plotting adsorption isotherms of test acids or bases, the Gibbs equation for adsorption [95] may be used:

$$\Gamma = -(1/RT)\,(d\gamma_{\ell s}/d\ln c) = (d\gamma_\ell \cos\theta/d\ln c)/RT, \tag{3.5}$$

where: c is the concentration of test acids or bases, Γ is the interfacial concentration. If adsorption isotherm is plotted by few points thus obtained, both the surface concentration of acidic and basic centers $\Gamma_м$ and the equilibrium constant of adsorption K_{eq} may be calculated graphically [79]:

$$c/\Gamma = 1/\Gamma_м K_{eq} + c/\Gamma_м. \tag{3.6}$$

If isotherms are determined for two or more temperatures, the molar adsorption heat of test acid or base is calculated from the van't Hoff ratio [17]:

$$\Delta H^{AB} = [RT_1T_2/(T_{2}-T_1)]\ln[(K_{eq}\ \text{при}\ T_1)/(K_{eq}\ \text{при}\ T_2)]. \tag{3.7}$$

This method can be used for determination of the Drago constants for polymers and for evaluation of the chemical nature of modifiers used for enhancing polymer adhesion [98].

3.9.3 *GRAPHICAL METHOD FOR DETERMINING SFE COMPONENTS*

Graphical method based on the Owens-Wendt equation [63] is a conventional and most commonly used method for obtaining the dispersive and acid-base SFE components. Basing on the equation $W_a = \gamma_\ell(1+\cos\theta) = W^d + {}^{AB} = 2(\gamma_s^d \gamma_\ell^d)^{1/2} + 2(\gamma_s^{AB} \gamma_\ell^{AB})^{1/2}$, the known γ_ℓ values of test liquids, and the measured contact angles on the test surface wetted by these liquids, the $(g_\ell^{AB}/g_\ell^d)^{1/2}$–$W_a/2(g_\ell^d)^{1/2}$ graph is plotted by linear approximation (where g_ℓ^{AB} and g_ℓ^d are the acid-base and dispersive SFE components of test liquids). The Y-intercept stands for the $(\gamma^d_s)^{1/2}$, and the slope of a line is equal to $(\gamma_s^{AB})^{1/2}$. The sum of the γ^d_s and γ_s^{AB} values thus obtained is the geometric mean approximation of the total SFE of material.

This method finds a wide application. Authors of this monograph measured contact angles on different polymer surfaces wetted by 11 test liquids among which there were neutral liquids (methylene iodide and α-bromonaphthalene), and soft and hard acids and bases (glycerol, dimethyl sulfoxide, phenol aqueous solution, and aniline) (Table 2.1).

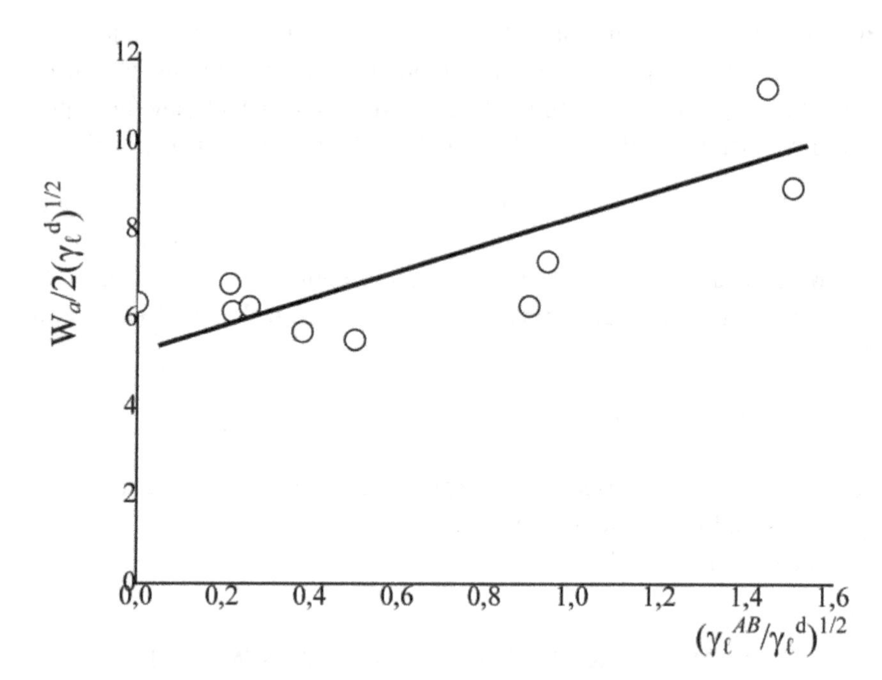

FIGURE 3.1 Graphical determination of SFE components of SEVA-20 coating.

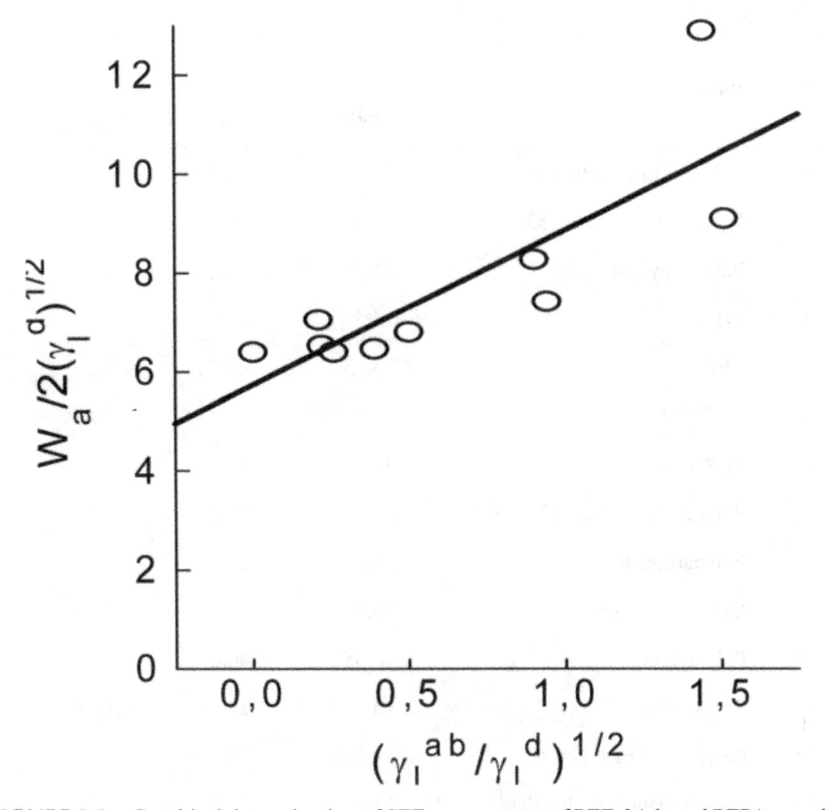

FIGURE 3.2 Graphical determination of SFE components of PEF-3A based PEPA-cured film.

Figures 3.1 and 3.2 visualize examples of the aforementioned graphs for two surfaces, namely SEVA-20 coating (Fig. 3.1) and PEF-3A based film (Fig. 3.2).

Tables 3.1–3.3 systematize the γ_s^d, γ_s^{AB} and γ_s values obtained by graphical method for a great number of polymer and metal samples. Tables involve virgin, or unmodified polymers and copolymers, unvulcanized rubbers purified by reprecipitation, modified polyolefins, epoxy compositions with different curing agents, as well as certain metal substrates.

As seen from Table 3.1, the lowest SFE value is observed for polytetrafluoroethylene, and the greatest one is observed for polypropylene carbonate, which is partly confirmed by data on thermodynamic work of adhesion of test liquids for these polymers (*see* Chapter 2, Table 2.2).

TABLE 3.1 Surface free energy and its components of different polymers.

No	Polymer	γ_s (mJ/m²)	γ_s^d (mJ/m²)	γ_s^{AB} (mJ/m²)
1	Polytetrafluoroethylene	25.2	25.1	0.1
2	Polyethylene grade 153	32.9	28.4	4.5
3	Polypropylene	34.25	33.25	1.0
4	EBA	34.5	28.4	6.1
5	EEA	35.35	29.05	6.3
6	SEVA 22	36.0	25.1	10.9
7	SEVA 29	38.6	25.0	13.6
8	Polyethylene grade 277–73	38.7	29.6	9.1
9	Polycarbonate	39.7	29.1	10.6
10	EVA terpolymer	40.4	27.9	12.5
11	Polystyrene	40.65	40.65	0
12	SEVA 14	40.7	29.4	11.3
13	Polymethyl methacrylate	41.25	38.9	2.35
14	Polyethylene terephthalate	43.7	35.5	8.2
15	Polypropylene carbonate	45.2	28.4	16.8

Acid-base SFE components of all polymers are different from zero, except for polystyrene. As mentioned above, the γ^{AB}_s depends on two values, namely the acidic and basic SFE parameters. Therefore, zero γ^{AB} can mean that either of these parameters or both of them are equal to zero in case of polystyrene, which is usually considered as almost neutral polymer. For other polymers, this value varies from 0.1 for polytetrafluoroethylene up to 16.8 for polypropylene carbonate. The same parameters of different rubbers are presented in Table 3.2.

TABLE 3.2 Surface free energy and its components of synthetic rubbers.

No	Rubber	γ_s (mJ/m²)	γ_s^d (mJ/m²)	γ_s^{AB} (mJ/m²)
1	IIR	24.9	20.6	4.3
2	EPDM	25.8	23.5	2.3
3	CIIR	27.9	23.3	4.6
4	BR	32.1	30.9	1.2
5	SKS-30	33.7	28.9	4.8
6	IR	34.4	28.9	5.5
7	Stereoregular BR	34.6	29.7	4.9

Compared with above mentioned polymers, the values of acid-base SFE components of rubbers are substantially less and fall in the range of 1.2 for butadiene rubber and 5.5 for isoprene rubber. It points to a low amphipathicity of rubbers, which seems as absolutely natural result due to their chemical nature.

TABLE 3.3 Surface free energy and its components of composite materials and metals.

No	Sample	γ_s^d, mJ/m²	γ_s^{AB}, mJ/m²	γ_s, mJ/m²
1	HDPE 273+DH	19.5	27.3	46.8
2	HDPE 273+PAT	18.4	28.2	46.6
3	PEF-3A+ DETA	34.4	7.9	42.3
4	PEF-3A+AF-2M	32.9	7.5	40.4
5	ED-20+DETA	36.1	6.8	42.9
6	ED-20+AF-2M	35.0	6.0	41.0
7	St3 steel	23.2	10.6	33.8
8	St20 steel	22.0	19.3	39.3
9	Brass	23.1	13.3	36.4

TABLE 3.3 *(Continued)*

No	Sample	γ_s^d, mJ/m²	γ_s^{AB}, mJ/m²	γ_s, mJ/m²
10	ED-20 (50%)+PEF-3A (50%) +PEPA	30.9	15.8	46.7
11	ED-20(54%)+PEF-3A (16%) +PEPA +15% talc + 15% chrome oxide	31.2	18.4	49.6

In addition, the SFE and its components were determined for a series of modified polymeric materials, depending on modifier content. Figures 3.3(a, b)–3.6 visualize the γ_s, γ_s^d and γ^{AB} curves for polyethylene coatings modified by primary aromatic amines (DH modifier and benzidine), ED-20-based polyepoxide modified by PEF-3A epoxy urethane rubber, and chlorobutyl rubber coatings modified by p-Quinone dioxime and manganese dioxide.

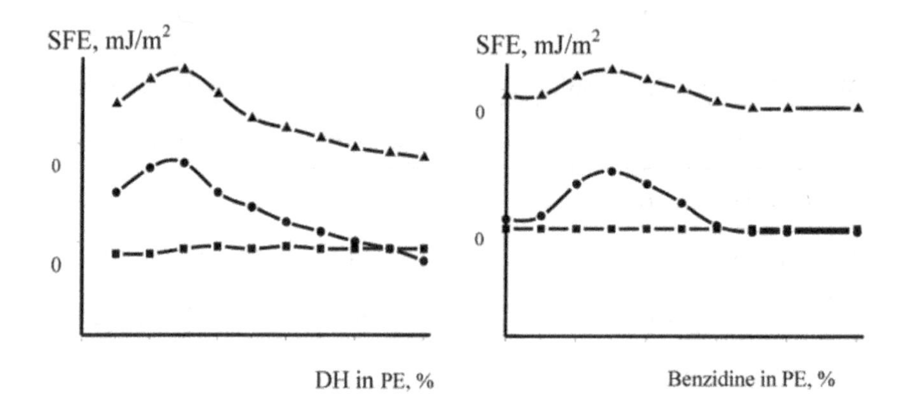

FIGURE 3.3 The dispersive SFE component (1), acid-base SFE component (2), and total SFE (3) of polyethylene coatings as a function of modifier content: (a) DH modifier; (b) Benzidine.

For polyethylene coatings, the acid-base component is increased by 10–15 mJ/m² in the concentration range of 1–3% in case of using both DH modifier (Fig. 3.3a, curve 2) and benzidine (Fig. 3.3b, curve 2). The dispersive component (curve 1) remains virtually unchanged in the whole concentration range to be near 20 mJ /m^2. Therefore, the total SFE γ_s (curve 3) also reaches the peak value in modifier content ranges stated. The increased γ_s is in a good agreement with data on the increased resistance to cathodic disbandment for coating [99]. As the resistance to cathodic disbandment is a characteristic of adhesion ability, it may be concluded that the increase in the γ_s occurs simultaneously with the enhancement of adhesive interaction in the systems in question.

Chemical modification of polyethylene resulting in polar amino groups in its structure, leads in its turn to the increased γ_s and W_a of the modified polyethylene and, hence, in the enhanced adhesive ability of this composite material due to intensification of interfacial interaction. At optimal amine concentrations in peak points shown in Fig. 3.3, the γ_s of coating comes close to the γ_s of substrate. This also goes to prove that adhesive interaction in these modified systems should increase to reach the maximal value at SFE extreme points.

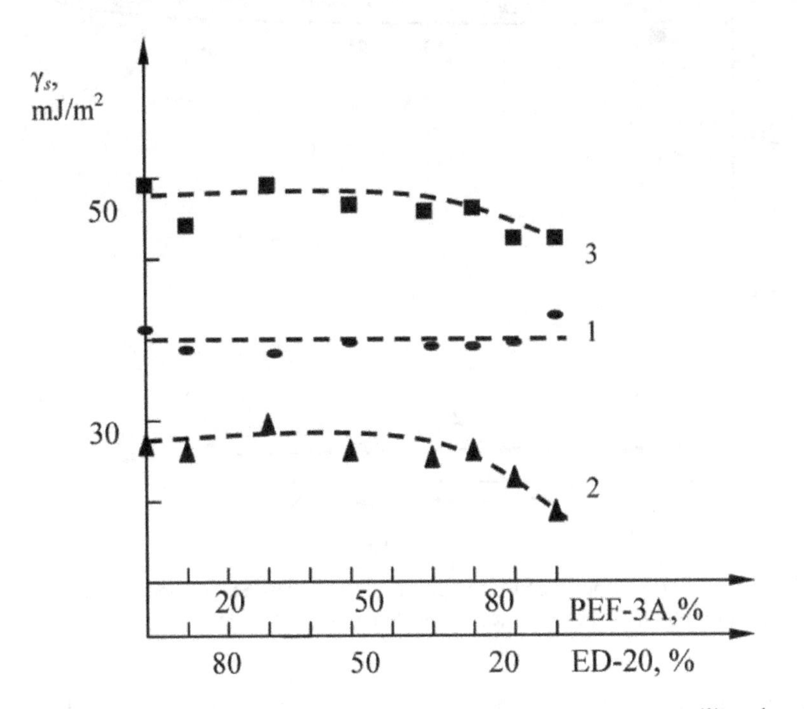

FIGURE 3.4 The dispersive SFE component (1), acid-base SFE component (2), and total SFE (3) of ED-20-based polyepoxide.

Total SFE of the cured epoxy primer is about 40 mJ/m^2, with the dispersive SFE component γ_s^d making a main contribution to this value (Fig. 3.4). As PEF-3A epoxy urethane rubber content rises, the dispersive component remains virtually unchanged, whereas both the acid-base component and total SFE values are decreased. This is explained by decreasing in concentration of the most polar OH-groups. For the virgin oligomer cured at the conversion rate of 0.7, the OH-group concentration is 14.7%, whereas for the virgin PEF-3A epoxy urethane rubber cured at the same conversion rate, the OH-group concentration is only 4.2% [100]. Thus, the SFE is a characteristic susceptible to variation in properties of adhesive due to introducing additives. This was verified for a great number of modified systems [99].

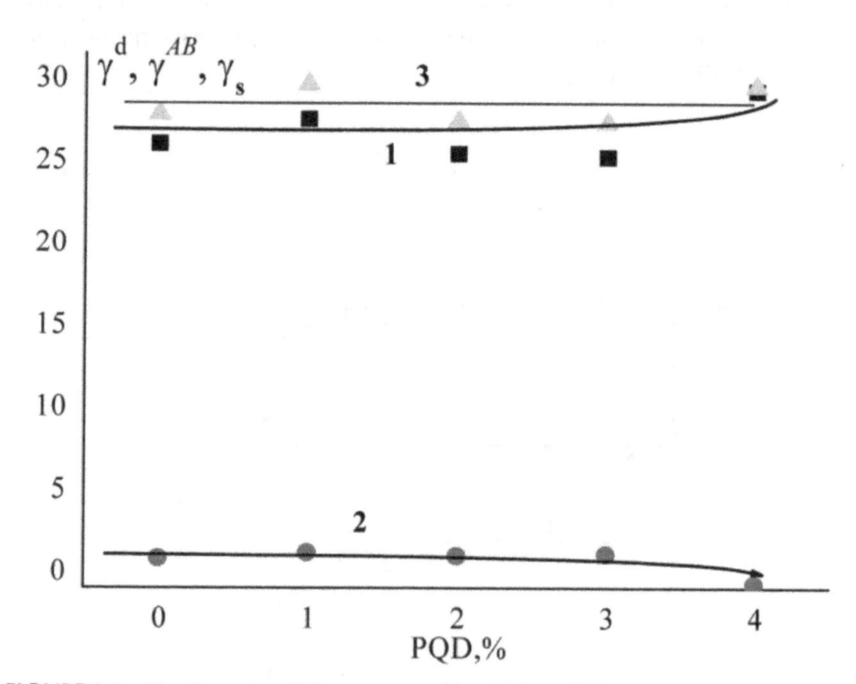

FIGURE 3.5 The dispersive SFE component (1), acid-base SFE component (2), and total SFE (3) of rubber-based composite as a function of PQD content.

For modified rubbers, no changes in values of the SFE and its components are observed in case of using both modifiers (Figs. 3.5 and 3.6).

Total SFE is near 30 mJ/m^2 in the presence of PQD and about 27–28 mJ/m^2 in the presence of manganese dioxide. In case of PQD, only slight increase in the γ^d and, simultaneously, the decrease in the γ^{ab} at 4% content are observed. It is well explained by the fact that excess unbound modifier exudes onto surface, with the SFE of PQD itself being possibly greater and self-association ability being less than those of composite. The acid-base component of rubber-based composite in both cases is very low in value, being about 2 and 5 mJ/m^2 correspondingly in case of using these two modifiers. These low values may be explained by low acidity (i.e., low γ^+ value) or low basicity (i.e., low γ^- value), or by combination of both factors.

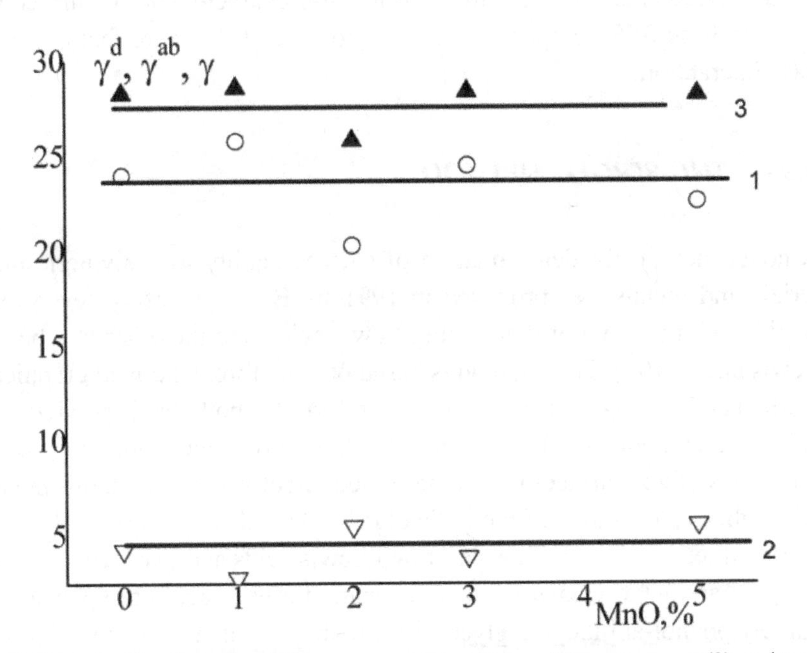

FIGURE 3.6 The dispersive SFE component (1), acid-base SFE component (2), and total SFE (3) of rubber-based composite as a function of manganese dioxide content.

Thus, using graphical method for determination of SFE components gives reliable and reasonable results. However, the knowledge of the dispersive and acid-base components is obviously insufficient for predicting functional abilities of a specific polymer or composite material.

At the same time, in case of rather complicated modified polymer systems, it is well known that the structure of surface and near-surface (and deeper) layers can differ. Distribution of chemically inactive modifiers follows diffusion laws and is not the same on surface and near support. Many modifiers as well as active functional groups of a high molecular weight polymer suffer any kind of steric hindrance on surface. All these factors cause the situation when polymer surface may not contain all acid-base centers existing in polymer matrix. The surface and "in-matrix" groups may differ by their ability to form acid-base bonds. Chemical composition of surface may be frequently unknown because of the aforementioned causes and lacking knowledge of all possible interactions between composite ingredients. That is why such integral characteristic of surface as the acid-base SFE component can help to predict the mere fact of acid-base interaction.

3.9.4 THE BERGER METHOD

A novel method for determination of surface acidity for polymeric materials and metals was proposed in 1991 by E. Berger using seven test liquids, with the two of them being Lewis acids and the other two being Lewis bases [101]. This method is based on the aforementioned graphical method and involves the geometric mean form for both the dispersive and acid-base components. Further modification allows the acidic and basic properties of test surfaces to be determined directly. First of all, this method implies calculations of the γ_s^{AB} values followed from individual interaction of test surfaces with each of two Lewis acids and two Lewis bases. The g_ℓ^{AB} and g_ℓ^{d} values are almost the same for each acid-base pair used, namely phenol-aniline and glycerol-formamide, which show a negligible difference in surface energies. The SFE of basic aniline is slightly greater than that of acidic phenol, and the SFE of basic formamide is slightly less than that of acidic glycerol. The base having the less SFE in one pair (formamide) should compensate the influence of the acid having the less SFE in another pair (phenol). If no acid-base interaction occurred, these pairs would have approximately equal contact angles on the test surface, and the γ_s^{AB} values determined for phenol-aniline and glycerol-formamide would be the same. But actually, it is not the case. The difference in the γ_s^{AB} values for acids and bases calculated according to the Eq. (3.8) gives us the measure of surface acidity, so the D value is referred to as the acidity parameter. The acidity parameter is widely used for estimating practical surface acidity values [102–111].

$$D = 2[(\gamma_s^{AB}\,(\text{aniline}))^{1/2} + (\gamma_s^{AB}(\text{formamide}))^{1/2}] - 2[(\gamma_s^{AB}\,(\text{phenol}))^{1/2} + (\gamma_s^{AB}\,(\text{glycerol}))^{1/2}] \quad (3.8)$$

The verification of the Berger method has been carried out for many polymers and polymer-based composite materials. The acidity parameters of unmodified polymers are contained in this section (Table 3.4). The results of method verification will be discussed later.

As said before, surface functionality can be hardly inferred by the γ_s^{AB} value. Unlike it, the acidity parameter gives us direct information of which properties, acidic or basic, a surface shows.

TABLE 3.4 Acidity parameters of unmodified polymers.

No	Sample	D, $(mJ/m^2)^{1/2}$	No	Sample	D, $(mJ/m^2)^{1/2}$
1	SEVA-22	–2.6	12	PPC	3.3
2	SEVA-14	–2.3	13	EVA terpolymer	4.1
3	Polystyrene	–1.8	14	Polycarbonate	7.0
4	HDPE	–1.1	15	IIR[1]	-1.5
5	PET	–0.85	16	EPDM	1.2
6	PMMA	–0.32	17	BR	1.5
7	PTFE	0.0	18	Stereoregular BR	2.0
8	Polypropylene	0.55	19	SKS-30	2.7
9	EEA terpolymer	1.8	20	IR	2.7
10	EBA	3.0	21	CIIR	3.6
11	LDPE	3.1	22	PVC	1.3

[1]Data for samples Nos. 15–21 correspond to rubbers purified from admixtures by reprecipitation. Rubber coatings were prepared from rubber solutions in organic solvent by casting onto a glass support.

For example, the γ_s^{AB} values for butyl rubber and chlorobutyl rubber are almost the same (4.3 and 4.6 mJ/m^2 correspondingly), but their acidity parameters differ noticeably. Butyl rubber is slightly basic (D = 1.5 (mJ/m^2)$^{1/2}$), whereas chlorobutyl rubber is acidic ((D = 3.6 (mJ/m^2)$^{1/2}$) due to chlorine atoms. By the acidity parameter, all unmodified rubber surfaces tested are rather neutral, which is stipulated by their low functional activity.

The γ_s^{AB} values for polycarbonate and SEVA-22 copolymer are also almost the same (10.6 and 10.9 correspondingly), but their acidity parameters differ profoundly. The polycarbonate surface has been found to be pronouncedly acidic (D = 7.0), which is probably caused by acidic bisphenol residues in its structure. At the same time, the SEVA-22 copolymer surface is basic (D = –2.6), since the vinyl acetate group contains the carbonyl group, which is basic due to greater electronegativity of the oxygen atom against the carbon atom.

By analyzing all the values shown in Table 3.4, it may be concluded that the Berger method allows true and chemically proved characterization of surface. According to all literature sources, polymethyl methacrylate is basic (D=0.32), but polyvinylchloride is acidic (D=1.3). The D=3.1 value for LDPE is the evidence of acidity due to thermal oxidation upon polymer molding, which means that acid-base properties are affected by conditions of surface preparation (molding time, temperature, etc.).

3.9.5 THE VAN OSS METHOD

The van Oss theoretical approach was discussed in Chapter 2 (Eqs. 2.28–2.30). For calculating the W_{12}^{AB}, the geometric mean remains but in a quite different form:

$$W_{12}^{AB}=2(\gamma_{1-}\gamma_2^{+})^{1/2} + 2(\gamma_1^{+}\gamma_{2-})^{1/2} \tag{3.9}$$

where: γ_{1-} is the basicity parameter of the phase 1, and γ_1^{+} is the acidity parameter of the phase 1. The same holds for the phase 2. Measurement units of these values are the same as used for the SFE. The values of acidic and basic parameters of any solid smooth surface can be determined from the calculated W_{12}^{AB} values by using test liquids with the known γ^- and γ^+. This method was further developed in detail in collaboration with Chaudhury and Good into the van-Oss-Chaudhury-Good (vOCG) theory, which has become the most commonly used abroad. The theory is based on the expression for thermodynamic work of adhesion for a liquid and a solid:

$$W_a = \gamma_\ell(1+\cos\theta) = 2(\gamma_\ell^{\ d}\gamma_S^{\ d})^{1/2} + 2(\gamma_\ell^{\ +}\gamma_{S-})^{1/2} + 2(\gamma_S^{\ +}\gamma_\ell^{\ -})^{1/2} \quad (3.10)$$

From the practical point of view, the Eq. (3.10) allows the dispersive, electron-donor, and electron-acceptor SFE components of solids to be calculated by using three test liquids as a minimum. It should be, however, noted that the results thus obtained are strongly dependent on the choice of these test liquids [97]. Determination of the electron-donor γ^- and electron-acceptor γ^+ parameters of test liquids themselves constitutes a serious problem. According to authors of the vOCG method, it cannot be applied without *a priori* γ^- and γ^+ values for any one of test liquids being of polar nature. Van Oss suggested water as such a liquid and assumed that it would equally show both acidic and basic properties at 20°C, and $\gamma_\ell^{\ -} = \gamma_\ell^{\ +} = \gamma_\ell^{\ AB}/2 = 25.4$ mJ/m^2. This assumption was not experimentally verified, and the values determined by van Oss are far from being perfect, since using those results in much lowered electron-acceptor parameter values for polymers.

3.9.5.1 THE DELLA VOLPE AND SIBONI MODIFICATION

The vOCG theory was modified by Della Volpe et al. by using matrix formalism and by writing equations in a matrix form [2, 97, 115–119].

$$\begin{matrix} \gamma_{l1}(1+\cos\theta_1)/2 \\ \gamma_{l2}(1+\cos\theta_2)/2 \\ \gamma_{l3}(1+\cos\theta_3)/2 \end{matrix} = A\begin{pmatrix} \sqrt{\gamma_s^{LW}} \\ \sqrt{\gamma_s^-} \\ \sqrt{\gamma_s^+} \end{pmatrix} \quad (3.11)$$

where: A is the 3rd order matrix obtained with the known γ_ℓ^d, γ_ℓ^+ and γ_ℓ^- values of test liquids.

The modified theory is about using only three and only polar liquids for calculating all SFE components, which substantially increases the error of the results obtained. Using more than three, correctly chosen liquids, which should be mainly acidic, basic, or dispersive is considered to be more correct. If more than three test liquids are used, and equations become overdetermined, the solution may be obtained by the best approximation method. The method implies such solution as to be suitable from the physical point of view. If there may be the only solution, it can be surely stated that the γ^d, γ^+ and γ^- values obtained are correct and useful for characterization of surface interactions. Authors revealed the three main problems arising from using the vOCG method and proposed the ways of solving those. The problems are as follows:

- All polymer surfaces tested by this method mainly show the basicity;
- Results are strongly dependent on the choice of the triplet of liquids being used for contact angle measurements;
- Square roots of acidic and basic parameters frequently have negative values that make no physical sense.

In the vOCG modified theory, Della Volpe and Siboni proposed novel SFE component values for water. As a result, the total increase in values of the acidic components and the decrease in those of the basic ones is observed upon solving the system.

Authors state that a great deal of work should be done in order to solve the problem of estimating SFE components, in particular, having in mind the choice of correct series of liquids representing all kinds of different liquids. Regarding negative values, the number of those decreases by many times if a large series of liquids checked is used, as well as the contact angles are interpreted correctly. In addition, negative values in many cases are much less than experimental error, so a real value may be equal to zero [115–119].

3.9.5.2 NONLINEAR DELLA VOLPE-SIBONI MODIFICATION

The linear matrix system (Eq. (3.11)) requires the knowledge of the acidic and basic SFE parameters of test liquids. As their numeric values available are unsatisfactory, Della Volpe et al. [118] proposed a nonlinear system (ℓ+s+ ℓs) of equations:

$$\gamma_{l,i} = \gamma_{l,i}^{d} + 2\sqrt{\gamma_{l,i}^{+} \times \gamma_{l,i}^{-}}, \forall i = 1,2,...,l$$

$$\gamma_{s,j} = \gamma_{s,j}^{d} + 2\sqrt{\gamma_{s,j}^{+} \times \gamma_{s,j}^{-}}, \forall j = 1,2,...,S \quad (3.12)$$

$$\gamma_{l,i}(1 + \cos\theta_{i,j}) = 2(\sqrt{\gamma_{l,i}^{d}\gamma_{s,j}^{d}} + \sqrt{\gamma_{l,i}^{+} \times \gamma_{s,j}^{-}} + \sqrt{\gamma_{l,i}^{-} \times \gamma_{s,j}^{+}}), \forall i,j$$

where: ℓ and s are the number of liquids and solids correspondingly. The total SFE $\gamma_{\ell i}$ values are reported in literature, the contact angles $\theta_{i,j}$ are determined experimentally. The first two equations follow from the Fowkes additivity concept [52]. Such a system has the advantage of being free of any *a priori* SFE component values. The total SFE values of test liquids determined by independent methods are thought to be known. Similar systems were solved many times (including this work) by using multi-objective optimization. However, the best approximate solution thus found demonstrated rather limited workability of this approach, since the calculated values of many parameters were not in agreement with chemical nature of sample.

Before discussing the vOCG method verification performed, it is necessary that a main prerequisite for this method, the equality of the acidic and basic parameters of water be considered.

3.9.5.3 SFE COMPONENTS OF WATER

An important role of SFE components of water if using the vOCG method was mentioned several times in previous sections. The values of those are considered as a unique starting point for determination of the acidic and basic parameters of other substances. According to van Oss, water equally shows both acidic and basic properties at 20°C. This assumption seems questionable enough, and many authors suggest other values of the acidic and basic parameters of water [97, 120–122].

In this concern, a great deal of work has been done by Abraham [120] who obtained the ratio $\gamma_\ell^+/\gamma_\ell^- = 5.5$ for water, which is substantially greater than 1 proposed by van Oss. According to Lee [121], this ratio is 3.2. A further, too high value $\gamma_\ell^+/\gamma_\ell^- = 42$ was proposed by Taft [122]. In other words, there is no common opinion on this point to be the subject of further considerations. For water at 20°C, the ratio $\gamma_\ell^+/\gamma_\ell^- > 1$ in the range of 3.2 to 5.5, according to [97], seems as the most acceptable. The equilibrium with atmospheric carbon dioxide in pure water is established quickly (in 10–20 min), with the pH-value being 5.5–6.0. This is an indication of a great amount of hydrogen ions, approximately 900 times greater compared with the OH-ions. Therefore, ordinary pure water may be said to be primarily acidic. In addition, it should be noted that there is no reported data on measurements of contact angles on surfaces of commonly used polymers wetted by extra-purified water (pH=7.0). For scientific applications, those measurements are of great interest, since water is in equilibrium with carbon dioxide in many systems, including biological ones. All mentioned before should be taken into account for explaining mainly basic nature of polymer surfaces estimated by van Oss [112, 113], probably due to the SFE minimization in the presence of atmospheric moisture at thermodynamic equilibrium. Though water always occurs in the environment and is always acidic, its amount is too low to play a substantial role in determining acid-base characteristics of polymer surface.

The attempt was made by Della Volpe and Siboni [97, 118] to clarify the problem concerning SFE components of water. They analyzed dependences of the acidic and basic parameters of 14 solid polymers and 10 liquids on the numeric value of the $\gamma_\ell^+/\gamma_\ell^-$ for water and obtained unexpected results. As the dispersive component of water reaches 23–24 mJ/m^2, a strong nonlinear deviation of parameters is observed for many polymers. For the basic parameter of DMSO, this nonlinear deviation occurs at 26–27 mJ/m^2. The conclusion was made thereunder that the best region for choosing the dispersive SFE component of water is 23–27 mJ/m^2. For mathematical optimization, Della Volpe and Siboni proposed the valuation function, by using the minimum of which the unknown components can be determined. This function has the minimum at $\gamma_\ell^d = 26.25$ mJ/m^2, which differs from the normally used 21.8 ±0.2 mJ/m^2. This widely used value was derived by Fowkes in 1964 from interfacial interactions of water with fully dispersive liquids. The analysis of data reported in literature [123–126] yields the average value of 27.1±5.7 mJ/m^2, which is also greater than that proposed by Fowkes. Of course, this might be concerned with the fact that liquid-liquid interaction differs from liquid-solid one even if interactions are fully dispersive. Thus, molecular motions and interactions between a solid polymer above the glass-transition temperature and water are more obstructed compared with those between water and other liquids. Moreover, the geometry of interactions is also various, since even the most smooth polymer surfaces differ from those of liquids.

Della Volpe and Siboni took the dispersive, acidic, and basic components of water as 26.25, 48.5, and 11.2 mJ/ m^2 correspondingly. They also took the $\gamma_\ell^+ / \gamma_\ell^-$ ratio as 4.35 that falls into the middle of the range of 3.2–5.5 considered earlier. Thus, according to their calculations, water happened to be a strong acid. Using the best approximation method and the values mentioned, they obtained more acceptable values of acidic and basic parameters of polymers and test liquids [97].

3.9.6 VERIFICATION OF THE VOCG METHOD FOR POLYMERS

As mentioned earlier, the SFE parameters of test liquids obtained by authors of the method must be corrected, because of being not corresponded sometimes to the chemical nature of substances. Results obtained by authors of this monograph regarding the vOCG method verification, taking into consideration and eliminating as much as possible all the drawbacks present, are given below. A special attention was paid to more accurate determination of the γ^d, γ^+ and γ^- of several test liquids, namely the most frequently used glycerol, formamide, and ethyleneglycol, basing on measurements of contact angles in systems involving paraffin and some polymers. By using the values obtained, SFE components and parameters of polymeric materials being used in adhesive techniques were then found.

Determination of the γ^d of test liquids was carried out by using neutral polymer surfaces, namely PTFE, paraffin, and 15303–003 and 15803–020 PE grades. It is assumed for those that $\gamma_S = \gamma_S^d$, but $\gamma^+ = \gamma^- = 0$. The γ_S^d for PTFE was calculated by wetting surface by a series of *n*-alkanes (such as

n-heptane, n-nonane, and n-decane) with the known values of surface tension according to the formula:

$$\gamma_s^d = \frac{\left[\gamma_l(1+\cos\theta)\right]^2}{4\gamma_l^d} = 21.07 mJ / m^2 \qquad (3.13)$$

The γ_S^d for paraffin calculated by using methylene iodide was 28.58 mJ/m^2. The dispersive SFE component values thus found for neutral surfaces were then applied for determining the dispersive SFE component of test liquids.

Further task was to determine the acidic and basic SFE parameters of test liquids. It was solved by using monofunctional (acidic or basic) polymer surfaces. PVC was used as conventional acidic polymer ($\gamma^- = 0$, $\gamma_S^d = 41.61$ mJ/m^2), and PMMA was used as the basic one ($\gamma^+ = 0$, $\gamma_S^d = 43.00$ mJ/m^2).

As is known [125], it is impossible to determine the γ^+ and γ^- parameters of liquids by measuring contact angles without taking these parameters for one polar liquid. Thus, water was used as such a liquid with commonly accepted initial parameters $\gamma_W^d = 21.8$ mJ/m^2 and $\gamma_W^+ = \gamma_{W-} = 25.5$ mJ/m^2.

Data obtained from calculations performed are presented in Table 3.5. The total SFE values for glycerol and formamide happened to be close to those reported in literature, namely 63.7 mJ/m^2 for glycerol and 57.5–58.4 mJ/m^2 for formamide according to different sources [101, 125, 126]. It should be also noted that SFE parameters obtained for test liquids are in qualitative agreement with those obtained by Lee [127] and Della Volpe and Siboni [97]. It can be seen that the acidic parameters increase, while the basic ones decrease, which better reflects the chemical nature of liquids. Particularly, this concerns glycerol, since the values of the basic parameter reported in world literature [53, 128] are usually substantially greater than those of the acidic one, which does not correspond to the chemical nature of glycerol considered to be a weak acid. The basic parameter of formamide also decreases down to 20.24 mJ /m^2 compared with that reported by van Oss (39.6). We obtained the basic parameter of formamide, which is close to that of later values obtained by Jan'czuk et al. [128].

TABLE 3.5 Total SFE and its components and parameters of test liquids (mJ/m²).

Liquid	γ_ℓ	γ_ℓ^d	γ_ℓ^+	γ_ℓ^-	γ_ℓ^{ab}
Water	72.8	21.8	25.5	25.5	51
Glycerol	63.61	34.18	16.56	13.08	29.43
Formamide	58.21	32.49	8.17	20.24	25.72
Ethyleneglycol	48.35	25.76	12.26	10.41	22.29
Methylene iodide	50.8	50.8	0	0	0

SFE components and parameters of test liquids being found, the unknown acidic and basic parameters of polymer surfaces can be determined. In simplest case, this task is solved by using the triplet of liquids. The system of three equations (3.11) for each surface according to the Della Volpe-Siboni modification was solved by using Cramer's rule and Microsoft Excel software application. A confirmation was obtained that the choice of test liquids for the triplet affects the results obtained [97, 115–119].

This fact was exemplified by SFE components and parameters of two polymer surfaces, PMMA and PVC, obtained by using different triplets and given in Table 3.6.

TABLE 3.6 SFE components and parameters of polymer surfaces (mJ/m²).

Surface	Triplet[1]	γ_S^d	γ_S^+	γ_{S-}	γ_S^{AB}	γ_S
PVC	MWF	41.61	1.81	0.10	0.85	42.46
	MGF	41.61	1.75	0.08	0.75	42.36
	MEF	41.61	1.48	0.01	0.24	41.85
	MWE	41.61	0.07	0.57	0.40	42.01
	MWG	41.61	1.46	0.03	0.42	42.03
	WGF	41.88	1.71	0.09	0.78	42.66
	WEF	39.24	4.71	0.94	4.21	43.45

TABLE 3.6 *(Continued)*

Surface	Triplet[1]	γ_S^d	γ_S^+	γ_{S-}	γ_S^{AB}	γ_S
PMMA	MWF	43.00	0.11	2.84	1.12	44.12
	MGF	43.00	0.28	1.86	1.44	44.44
	MEF	43.00	0.03	3.80	0.68	43.68
	MWE	43.00	1.14	9.48	6.57	49.57
	MWG	43.00	2.51	0.18	1.34	44.34
	WGF	40.48	0.47	2.25	1.06	44.06
	WEF	45.08	0.00	3.55	0	45.08

[1]Triplet abbreviations: M is methylene iodide, W is water, F is formamide, G is glycerol, and E is ethyleneglycol.

As shown from Table 3.6, some of triplets show almost the same values, except for WEF (water-ethyleneglycol-formamide) triplet for PVC, and both WEF and MWE (methylene iodide-water-ethyleneglycol) triplets for PMMA. The problem of correct choice of the triplet and estimation of its workability is a focal point for researchers since the moment of the vOCG theory appearance, and there are only uncompiled recommendations on this point so far. Therefore, we will enlarge upon the question of triplet choice.

It is also evident from the table that the acidic component of PVC shows slightly greater values compared with PMMA, while the latter is greater in the basic component values. This is in agreement with the chemical nature of these polymers.

3.9.6.1 SCIENTIFICALLY REASONED APPROACH TO TRIPLET CHOICE

To eliminate systematic errors and to simulate different interactions of polymer surface with test liquid, all kinds of liquids should be used in the same triplet: fully dispersive, mainly or fully acidic, and mainly or fully basic liquids [97]. Using more than three liquids also reduces estimation error, but mathematical tool gets more complicated, since the system of equations becomes over determined, and more than one solution is possible.

Della Volpe exemplified the importance of correct triplet choice by using triangular diagrams, where the bottom, left, or right sides correspond to the values of the normalized dispersive component of the liquid as well as acidic and basic parameters respectively. Normalization implies dividing a component or parameter by the total SFE value. A triplet is denoted by a certain symbol (square, delta, etc.) located in the center of mass of the corresponding triangle. The triangular diagram of the triplets of liquids based on values obtained by van Oss and Good clearly demonstrates that the basic components prevail, since the majority of points fall near the right side of triangle. The acidic SFE parameters of virtually all test liquids are very low. The acidic parameters calculated by using the Della Volpe-Siboni modification are greater, and symbols corresponding to triplets are better distributed throughout triangular diagram.

In contrast to diagrams mentioned, we plotted three-dimensional triangular diagrams corresponding to the re-calculated triplets (Table 3.5). According to our interpretation, each triplet is characterized by a triangle in three-dimensional matrix (Fig. 3.7). This seems to be more reasoned, since the γ^d, γ^+ and γ^- values are not interdependent and plotted each along its axis, with their normalization being not necessary.

It is evident that in order for different interactions to be simulated, test liquids used should have different functionality. As a result, the longer the distance between points in corners of triangles, the more liquids differ by their nature and, hence, the more the area of a triangle formed therefrom. Therefore, triangle area may serve as a criterion for triplet workability. Figure 3.7 depicts two triangles for MWE and GEF triplets having noticeably differing areas. Evidently, MWE triplet as a triplet having greater area is more suitable for solving the problems in question, compared with GEF triplet containing no dispersive liquid.

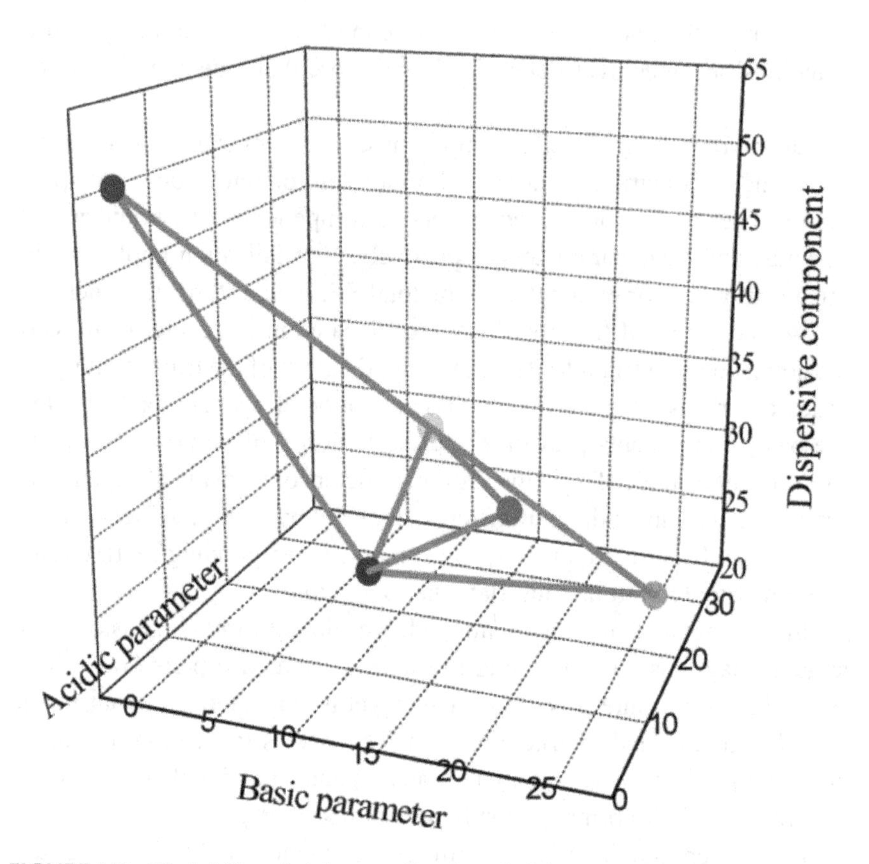

FIGURE 3.7 Workability of triplets as a function of triplet area.

Authors [97] suggest checking suitability of a triplet by calculating its correlation coefficient K_n. The K_n depends on the matrix norm A which is defined by so called vector norm and denotes a common error in wetting data:

$$K_n = ||A||_1 \cdot ||A||^{-1}, \tag{3.14}$$

where: $||A||_1$ is the matrix norm to be calculated as a maximal sum of elements of j-th column, $||A||^{-1}$ is the inverse matrix norm. A great K_n value means that the triplet of liquids shows a great measurement error.

The correlation coefficients calculated by authors for seven triplets of liquids as well as triangle areas of the corresponding triplets shown in triangular diagrams are presented in Table 3.7.

TABLE 3.7 Correlation coefficients of triplets and triangle areas.

Triplet	MWF	MGF	MEF	MWE	MWG	FWG	FWE
Correlation coefficient	17.44	15.27	28.35	106.43	60.86	14.15	26.98
Triangle area, units^2	315	253	240	35	70	121.5	108

By analyzing the table, it may be concluded that both methods for evaluation of triplet workability are in a good agreement with each other. However, the correlation coefficient determination requires complicated mathematical calculations. It should be noted that using many triplets is limited because of negative roots occurred in some cases, which makes no physical sense.

Using the corrected vOCG method allowed SFE components as well as the acidic and basic parameters of different polymer and inorganic surfaces to be obtained (Table 3.8).

TABLE 3.8 Surface-energy characteristics of different samples (mJ/m^2).

Surface	γ_S^d	γ_S^+	γ_{S-}	γ_S^{ab}	γ_S
PMMA	43.0	0.14	2.83	1.08	44.08
PVC	41.61	1.68	0.06	0.61	42.22
PP	30.12	1.86	0.31	1.52	31.64
PET	36.82	3.2	0.37	2.2	39.02
EVA	36.27	0.36	6.37	3.03	39.3
St3 steel	32.51	8.78	0	0	32.51

As seen from the table, PMMA and EVA are mainly basic, while PVC and steel are acidic. In case of polymers, the data obtained are in agreement with chemical nature of polymers, whereas the acidic nature of steel is due to oxides forming on its surface that will be discussed later. Thus, the vOCG method verification performed confirmed its workability and applicability to examination of acidic and basic properties of polymer and inorganic materials.

3.9.6.2 VERIFICATION OF THE NONLINEAR VOCG METHOD MODIFICATION

The Eq. (3.12) is solved provided that sufficient number of liquids and solid surfaces are used. For example, in case of ℓ=7 and s=4 the system consists of 39 equations and 37 unknowns. However, the resultant solution for certain substances is far enough from being true. This problem gives rise to correction of nonlinear vOCG method, which has been performed by authors of this monograph [129, 130]. The Eq. (3.12) was simplified by reducing the number of unknowns due to the dispersive SFE components γ^d of liquids and polymers taken from reliable sources [3, 125]. Polymers such as EVA, PTFE, PMMA, and PET were used as solid surfaces. Further solution also contained "failed" parameters and depended profoundly on variation of starting conditions. This absolutely mathematical method of solution is most likely to be unsuitable for a given problem. In each case of changing the order of writing equations the results will differ, since the calculation algorithm (the gradient descent method) will use different starting parameters. In our case, calculation of the acidic and basic parameters of substances as physically independent values is performed not simultaneously but in turn.

In spite of drawbacks mentioned, the ranges of the starting values where unknowns for polymers have the fixed values varying in a small region near the equilibrium happened to exist. Components and parameters of polymers slightly "respond" to changes in the starting conditions to be virtually unchanged. This is probably due to mathematically smooth objective function in the corresponding scan regions. Results being thus obtained characterize the chemical nature of test surfaces well enough. After multiple solving the system by slight varying the starting conditions the averaged values of SFE components and parameters of polymers were obtained, presented in Table 3.9.

TABLE 3.9 SFE components and parameters of polymers.

Polymer	γ^d	γ^+	γ^-	γ^{ab}	γ
SEVA-14	30.76	0.06	0.21	0.22	30.98
PTFE	29.84	0.00	0.00	0.00	29.84
PMMA	40.39	0.03	1.47	0.40	40.79
PET	39.47	0.00	3.26	0.00	39.47

Should these polymer surfaces be now considered as test ones, the problem is substantially simplified due to independent equations occurred in the system. The initial system of equations ultimately falls into 7 subsystems (since 7 test liquids were used) each containing 4 equations and two unknowns, that is, the acidic and basic parameters of each liquid.

This two-variable optimization does not constitute serious problems in root determination, and we obtained the acidic and basic SFE parameters of liquids, presented in Table 3.10.

TABLE 3.10 SFE components of test liquids.

Liquid	γ^+	γ^-
Glycerol	16.38	14
Methylene iodide	1.32	0
α-Bromonaphthalene	0.05	0
Formamide	5.25	39.89
Water	61.45	10.58
Dimethylsulfoxide	1.25	14
Aniline	0.02	43.91

The "new" parameters differ from those reported in literature, well correspond to the chemical nature of test liquids, and give consistent results in case of determining properties of the unknown surfaces. The data for aniline are presented for the first time. Greater basic parameters are observed for aniline, formamide, and dimethylsulfoxide, according to their chemical nature, whereas greater acidic parameter is observed for glycerol. The confirmation of up-to-date scientific views that water shows acidic properties (due to absorption of CO_2 from air) was obtained. The dispersive liquids (α-bromonaphthalene and methylene iodide), in accord with the data obtained and recent data reported in literature [54, 128], are not considered to be fully dispersive. Verification results for the obtained parameters of different polymer surfaces are presented in Table 3.11.

TABLE 3.11 SFE components and parameters of polymer surfaces.

Polymer	γ	γ^d	γ^+	γ^-
Polyvinylchloride	39.0	37.75	1.25	0.3
Polycarbonate	40.3	39.0	2.2	0.19
Polystyrene	41.2	41.2	0	0.06
Polypropylene	34.16	34.15	0.95	~0
Butyl rubber	22.3	20.9	0.67	0.7
ED-20+DETA	39.05	32.4	7.53	1.47
Polyethylene	36.46	36.46	0	0.5

Thus, polystyrene, butyl rubber, and polyethylene are neutral, while polypropylene, polyvinylchloride, and polycarbonate are slightly acidic, with the acidity increasing from polypropylene to polycarbonate. Polyepoxide surface is mainly acidic due to mobile hydroxyl groups.

Data obtained by using the last method (Tables 3.9–3.11) differ from those obtained by solving linear systems of equations (Tables 3.5–3.8). That makes sense because mathematical solution algorithms are different, but experimental data remain unchanged. In addition, the result obtained by solving nonlinear systems of equations seems as more reliable and consistent.

Using the data thus obtained allows correct estimation of SFE parameters and components of liquids, polymers, as well as other solid smooth surfaces, such as metals. This is significant for achieving better interfacial interaction to choose adhesive joint components, for targeted modification and choosing conditions of adhesive joint formation.

3.9.7 PRACTICAL APPLICATION OF ACID-BASE CHARACTERISTICS

To summarize the above information it can be said that the described techniques for estimation of acid-base properties yield different quantitative characteristics needed for further analysis, such as the Drago constants, the Gutmann donor and acceptor numbers, the acidity parameter, acidic and basic SFE parameters, heats of wetting and adsorption, etc. By using those, the effect of targeted modification or preliminary surface treatment prior to adhesive joint formation can be estimated to enhance adhesive properties, as well as the adhesive ability may be predicted. The comparative analysis, however, showed that not all of those characteristics are versatile in use, with some of them being less informative.

For example, the Gutmann donor and acceptor numbers D_N and A_N, proposed for liquids, allow only a single-valued assessment of acidity or basicity, with both electrostatic and covalent contributions being not considered. These numbers have different scales and measuring units and are determined in different ways, by spectroscopy or calorimetry, the methods which are both limitedly applied to polymers and less applicable to polymeric coatings. It should be noted that Drago found the enthalpy of formation of acid-base systems in gaseous phase and in pure solvents, therefore a large series of values obtained by Drago is also not applicable to solid polymer surfaces. Moreover, the Drago method allows the assessment of either acidity or basicity, with amphoteric properties being out of the question. The drawback of many up-to-date examinations of acid-base interactions is that researchers usually consider only the acidity of model acids or basicity of model bases, not taking into account bipolarity virtually each substance possess to some extent. For example, all polymers except for saturated hydrocarbons have either acidic or basic properties. Polyesters and PMMA are mainly basic, but also show noticeable acidic properties and are able to some self-association in solutions. Similarly, PVC is mainly acidic due to methylene chloride, but is prone to some self-association due to a weak but existing basicity of chlorine atoms [1].

Differential calorimetry is also not versatile in use. It may be used for examination of powdered suspensions only if particles flocculate in a stirred calorimeter vessel not too rapidly and have a surface area not less than 20 m^2/g for providing sufficient sensitivity. Differential calorimetry is not applicable for studying iron oxides because of magnetic interaction.

To determine the γ_s^d and acidity or basicity of powdered polymers, inverse gas chromatography finds expanding application [131–135]. The cases of applying it even to the mixture of modified polymers are reported [136]. Inverse gas chromatography is limited in use, because disintegration of test surface is needed, which is not always feasible.

Acid-base properties can be determined by ellipsometry provided that a surface is smooth and reflects well enough.

NMR chemical shifts can also provide information about heats of acid-base interaction in calculating the Drago constants and the A_N. However, all measurements are carried out in solutions rather than in real adhesive compositions. For solid smooth polymers, wetting measurements are preferred.

Among the existing and discussed wetting techniques, no one can be considered to be "ideal" or free of drawbacks. Nevertheless, some of them seem as more accessible and informative for solving specific practical problems. Drawbacks of the Fowkes-Mostafa method have been already mentioned. Even if we neglect theoretical errors of this method, its practical application will require molar heats of acid-base interaction of two interacting substances. Again, the less applicable to polymeric coatings calorimetry should be used. In addition, the interfacial concentration of acid-base bonds n_{AB} may be calculated only for the surface having completely known composition, which is not possible in many cases.

In principle, the vOCG theory is a good tool for estimating acid-base properties of polymer surfaces and interfacial interactions for possible further surface modification and tailoring, but the problem of practical realization of these prospects is far from being solved. Results being obtained are strongly dependent on the triplet chosen, which is the major drawback of this method. In some cases, square roots of acidic or basic parameters have negative value, which makes no sense. In spite of drawbacks mentioned, it is the vOCG theory (expressed as a nonlinear Della Volpe-Siboni modification) that has most proponents [1, 137–145]. This method is one of the most fundamental and in-demand and allows more detailed information about the energy of polymer surfaces. To solve problems, which arise, more test liquids and, hence, more correct triplets are required. A nonlinear modification of this method is under development and seems as a promising way of solving adhesion-related problems in spite of being mathematically complicated.

Regarding the Berger method, in spite of criticism of writing the acid-base components as the geometric mean Owens-Wendt equation, it is this equation that is used by many researchers for solid-liquid systems. This is explained by its relative simplicity and, what is more important, correct data yielded from this method. The acidity parameter provides information about acid-base properties of test surface and, hence, about the ability to take part in acid-base interaction with substrate. It allows estimating the integral characteristic of the acidity or basicity of any solid surface, including the end product, whereas most methods are only suitable for estimating properties of individual substances in the form of liquid or powder.

The review presented shows that in order to determine acid-base properties of organic and inorganic solid surfaces, it is better to choose methods in accord with contact angle estimation, among which both the Berger and vOCG methods are preferred. Next chapter shows that these methods make the control of modification of polymers as well as process conditions and coating preparation ways possible.

KEYWORDS

- **ellipsometry**
- **inverse gas chromatography**
- **IR and NMR spectroscopy**
- **microcalorimetry**
- **van Oss Method**

CHAPTER 4

TARGETED MODIFICATION OF ACID-BASE PROPERTIES OF POLYMERIC COMPOSITE MATERIALS

CONTENTS

Physical-chemical modification intended for improving different performance characteristics of polymer coatings is also able to have a targeted effect on their acid-base properties.

This chapter involves the ways of modification easily feasible in practice for possible further control of acid-base properties of a great series of composite materials, namely polyepoxide, polyolefin, rubber, and metal substrates. In general, acid-base properties of all surfaces were estimated according to the Berger method [101]. For some polyepoxides and EVA-based compositions, the concurrent examinations were carried out in accord with the vOCG method [112–114].

4.1 ACID-BASE PROPERTIES OF POLYEPOXIDES

Epoxy compositions widely used in production of glues, encapsulating and impregnating electroinsulation materials, and protective coatings for equipment to operate in aggressive environments are something special due to their unique properties, first thanks to excellent adhesion to different supports. By controlling formulation as well as conditions under which epoxy composition is produced, the range of processing and performance characteristics of coatings can be substantially improved. Targeted modification of surface properties intended for enhancing acid-base interaction in adhesive joint may facilitate improving coating quality and lifetime. In this concern, the examination of acid-base surface properties (SFE components and the acidity parameter) of the systems based on epoxy oligomers, as well as the effect of modification on these properties are of great interest.

We shall consider SFE components and the acidity parameters of solid smooth polyepoxides affected by different factors.

4.1.1 THE EFFECT OF CURING METHOD ON ACID-BASE PROPERTIES OF POLYEPOXIDES

A comparative estimation of the acidity parameters of cold cure (air-cured at room temperature) and hot cure (exposed to temperatures of 80–100°C over 2 hours) epoxy primers showed that the way of curing substantially affects this characteristic (Table 4.1). The SFE of hot cure samples is slightly less than that of cold cure ones, with both the dispersive and acid-base components decreasing. This possibly occurs because a curing reaction proceeds slowly at lower temperatures, and an amine reacts with atmospheric moisture and carbon dioxide to form ammonium salt, which is a Lewis acid.

TABLE 4.1 Acidity parameters of polyepoxides cured by different methods.

Sample	Acidity parameter D $(mJ/m^2)^{1/2}$	
	Cold cure	Hot cure
ED-20 + DTB-2	2.0	0.5
ED-20 + DETA	6.8	4.5
PEF-3A + DETA	4.0	3.0
PEF-3A + AF-2M	1.1	-0.5
PEF-3A + PEPA	3.8	2.8

The fact revealed is not unexpected, since even Berger stated [101] that changing the way of surface preparation and treatment (such as selecting a solvent for ungreasing) causes the variation in all surface characteristics of material.

4.1.2 THE EFFECT OF CURING AGENT ON ACID-BASE PROPERTIES OF POLYEPOXIDES

4.1.2.1 USING THE BERGER METHOD

Studying the effect of curing agent on the acid-base properties of polyepoxides is of great interest. By varying epoxy formulation, that is, by changing curing agent, the range of processing and performance characteristics of epoxy-amine systems can be substantially improved [146–156]. Conventional curing agents showing high to moderate to low activity allow targeted control of strength, dielectric, and thermal-physical characteristics of epoxy materials. It is known that the adhesive ability of coating decreases as both temperature and service life rise. In this concern, of great interest is studying the effect of the structure of curing agents on chemical reactions and curing pattern as well as on adhesive and surface-energy properties of epoxy-amine systems.

Acid-base surface characteristics of a series of epoxy systems cured by different curing agents are presented in Tables 4.2–4.3.

TABLE 4.2 The SFE and acidity parameters of epoxy primers cured by different curing agents.

Sample	γ_s^d, mJ/m^2	γ_s^{ab}, mJ/m^2	γ_s, mJ/m^2	D, (mJ/m^2)$^{1/2}$
ED-20 + DETA	36.10	6.80	42.90	6.80
ED-20 + DTB-2	31.20	6.90	38.10	2.00
ED-20 + AF-2M	32.50	7.40	39.90	3.40
ED-20 + PEPA	31.70	16.50	48.20	5.10
ED-20 + CS # 1[1]	32.60	11.70	44.70	–0.70
ED-20 + CS # 2	32.60	10.30	42.90	–0.60
ED-20 + CS # 3	30.90	12.30	43.20	–0.50
ED-20 + CS # 4	32.40	8.75	40.15	0.35
ED-20 + CS # 5	34.60	8.15	42.75	–1.65
ED-20 + CS # 6	33.30	9.80	43.20	–0.95
ED-20 + CS # 7	31.10	13.10	44.20	0.60
ED-20 + CS # 8	35.20	9.70	44.90	2.60
ED-20 + CS # 9	31.20	15.60	46.80	2.85
ED-20 + Kroot-1	31.60	6.15	37.75	-2.30
ED-20 + UP-583	25.40	31.90	57.30	6.80

TABLE 4.2 *(Continued)*

Sample	γ_s^d, mJ/m^2	γ_s^{ab}, mJ/m^2	γ_s, mJ/m^2	D, (mJ/m^2)$^{1/2}$
PEF-3A + DETA	34.40	7.90	42.30	4.00
PEF-3A + PEPA	34.00	8.20	42.20	3.80
PEF-3A+AF-2M	32.80	8.00	40.80	1.10

[1]CS #1–CS #9 are curing systems comprising PEPA and compounds presented in Table 4.4 in 50:50 mole ratio.

Curing systems CS #1–CS #9 and complex compounds based on Lewis acids and tri(halogenalkyl)phosphates (Tables 4.4 and 4.5) were examined as new generation curing agents.

It follows from Tables 4.2 and 4.3 that the nature of curing agent substantially affects sample surface.

The acidity parameter D of primers based on ED-20 epoxy resin and PEF-3A epoxy urethane rubber in combination with conventional curing agents has positive values and ranges from 1.1 to 6.8 (mJ/m^2)$^{1/2}$, which is the evidence of the acidic nature of surfaces. The more noticeable D value is observed for the systems ED-20 + DETA (6.8), ED-20 + UP-583 (6.8), and ED-20 + PEPA (5.1). This can be explained by the presence of functional groups on the surface of composition, which may serve as acid-base centers. These groups include OH-groups and >N̈ – groups (for ED-20-based compositions) and OH-groups, urethane groups, and –Ö-groups (for epoxy urethane compositions) which can be divided (according to Lewis) into two types—acidic (OH-groups and urethane groups) and basic (>N̈ – and –Ö-groups). The epoxy oligomer has the prevailing OH-groups and >N̈ -groups, with the OH-groups being more mobile and the surface being therefore mainly acidic. PEF-3A epoxy urethane rubber has urethane groups as acidic centers, with the surface OH-group content decreased. Simultaneously, the basic –Ö-groups are present on surface in greater amount during the curing. This affects surface properties, and the acidic nature of the PEF-3A-based coating is less pronounced.

TABLE 4.3 The SFE and acidity parameters of epoxy compositions cured by complex compounds based on Lewis acids and tri(halogen) phosphates.

Sample	γ_s^{ab}, mJ/m²	γ_s^{d}, mJ/m²	γ_s, mJ/m²	D, (mJ/m²)$^{1/2}$
1	3.25	36.8	40.05	4.55
2	1.7	39.45	41.15	4.3
3	8.5	33.75	42.25	1.3
4	3.3	36.5	39.8	2.85
5	2.35	38.7	41.05	2.7
6	9.45	30.4	39.85	4.25
7	6.3	33.7	40.0	4.3
8	5.6	33.35	39.75	3.7
9	5.6	36.2	41.8	3.4
10	5.8	33.65	39.35	2.8
11	2.15	35.45	37.6	-0.08
12	3.95	33.95	37.8	3.15

TABLE 4.4 Curing systems CS #1–CS #9.

No	Name
CS # 1	N, N'-di(3-phenoxy-2-hydroxypropyl) ethylenediamine
CS # 2	1,6- N, N'-di(3-phenoxy-2-hydroxypropyl) hexamethylenediamine
CS # 3	N, N'-di(3-allyloxy-2-hydroxypropyl) Ethylenediamine
CS # 4	1,6-N, N'-di(3-allyloxy-2-hydroxypropyl) hexamethylenediamine
CS # 5	N-(3-phenoxy-2-hydroxypropyl) aminoethanol
CS # 6	N-4-(2,3-hydroxypropyl) aminobenzoic acid
CS # 7	N-4-(3-hydroxy-2-chloropropyl) aminobenzoic Acid
CS # 8	N, N'-di(3-chloro-2-hydroxypropyl) ethylenediamine
CS # 9	Aminobenzylaniline

Different situation takes place in case of ED-20 + CS #1–CS #9 + Kroot-1 formulations. Most of them are slightly basic, and using the Kroot-1 curing agent makes the surface of epoxy coating pronouncedly basic [157].

TABLE 4.5 Curing systems composed of complex compounds based on Lewis acids and tri(halogenalkyl)phosphates.

Sample	Curing system
1	25% solution of bis[tri(2-chloroethyl)phosphate] tetrachlorotin in tri(2-chloroethyl) phosphate
2	8.3% solution of bis[tri(2-chloroethyl)phosphate] tetrachlorotin in tri(2-chloroethyl) phosphate
3	8.3% solution of bis[tri(2-chloropropyl)phosphate] tetrachlorotin in tri(2-chloropropyl) phosphate
4	25% solution of bis(tributylphosphate) tetrachlorotin in tributylphosphate
5	8.3% solution of bis(tributylphosphate) tetrachlorotin in tributylphosphate
6	50% solution of bis[tri(2-chloroethyl)phosphate] pentachloroniobium in tri(2-chloroethyl) phosphate
7	Bis[tri(2- chloroethyl) phosphate] tetrachlorotitanium
8	Bis(tributylphosphate) tetrachlorotitanium
9	Bis[tri(2- chloropropyl) phosphate] tetrachlorotitanium
10	Bis[tri(2- chloroethyl) phosphate]dichlorozinc
11	Bis[tri(2- chloropropyl) phosphate] dichlorozinc
12	Bis(tributylphosphate) dichlorozinc

The feature of complex compounds based on Lewis acids and tri(halogenalkyl)phosphates is that polymerization in the presence of proton donors proceeds under the influence of a complex acid $H_2[(RO)_2{:}MCl_4]$ (where M is Sn, Ti, or Zn) which results from the reaction of $2P{\times}MCl_4$

(where P is a phosphate molecule) and compounds containing proton donor groups (water, alcohols, etc.) always present in ED-20 epoxy resin:

$$2P\cdot MCl_4 \xrightarrow[-2F]{2ROH} H_2\left[(RO)_2 : MCl_4\right] \Leftrightarrow 2H^+ + \left[(RO)_2 : MCl_4\right]^-, \quad 4.1$$

where R is H or alcohol residue.

Further, homopolymerization of ED-20 epoxy oligomer occurs to yield polyethers [146]:

$$R-HC\underset{O}{\overset{}{\diagdown\!\diagup}}CH_2 \xrightarrow{H^+} R-\underset{OH}{CH}-\overset{+}{C}H_2\cdot \xrightarrow{nR-HC\underset{O}{\diagdown\!\diagup}CH_2} -\!\!(\underset{R}{CH}-CH_2-O)_{\overline{n}}, \quad 4.2$$

where R is ED-20 epoxy oligomer residue:

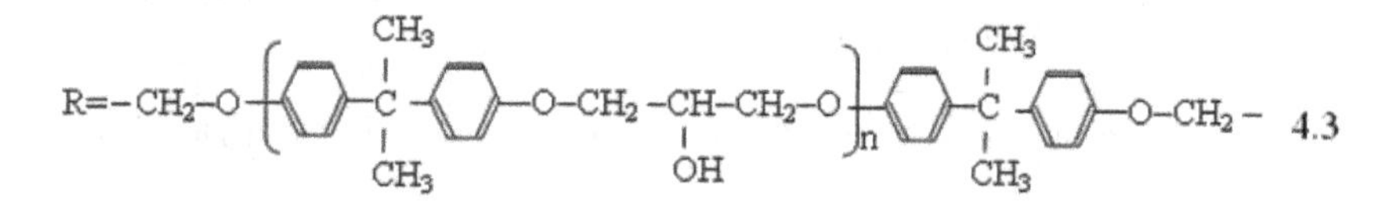

In this case, the nature of curing agent also noticeably affects surface characteristics of samples. The acidity parameter D is somewhat less than in case of conventional curing agents also showing, however, positive values and ranging from –0.08 to 4.55 $(mJ/m^2)^{1/2}$, which is the evidence of mostly acidic nature of the surface of polyepoxide coatings using new generation curing agents.

The acid-base SFE component of this series of epoxy samples is relatively not too high and ranges from 1.7 mJ/m^2 for the sample # 2 to 9.45 mJ/m^2 for the sample # 6. The dispersive SFE component ranges from 30.4 mJ/m^2 for the sample # 6 to 39.45 mJ/m^2 for the sample # 2. Without going into the majority of chemical reactions occurring upon curing of all samples tested, we will only state the fact that choosing curing agent allows targeted achievement of acidic or basic properties of epoxy coatings in rather broad ranges [111, 149–156].

4.1.2.2 USING THE VOCG METHOD

In order to compare estimations obtained by Berger and van Oss, the determination of the acidic γ_s^+ and basic γ_{s-} SFE parameters of polyepoxide coatings cured by different curing agents is of interest. To minimize the effect of the choice of test liquids on resultant values, it is more preferred to use a nonlinear modification of the vOCG method using parameters of test liquids presented in Table 3.9 (Chapter 3, Section 3.9.6.2). Epoxy coatings cured by complex compounds based on Lewis acids and tri(halogen) phosphates, namely those showing differences in acid-base properties estimated by the Berger method [158–160]. Thus, the most acidic samples # 1, 2, and 6; slightly acidic, almost neutral sample # 3, and slightly basic, almost neutral sample # 11 were chosen (Table 4.3). The results can be seen in Table 4.6.

TABLE 4.6 Acidic, basic, and dispersive characteristics of epoxy samples.

Sample	γ_s, mJ/m^2	γ_s^d, mJ/m^2	γ_s^+, mJ/m^2	γ_{s-}, mJ/m^2	γ_s^{ab}, mJ/m^2
1	41.7	41.7	1.5	0	0
2	46.2	43.85	2.23	0.61	2.33
3	41.3	41.3	0	0.26	0
6	36.5	34.95	2.48	0.24	1.54
11	33.3	32.0	0.21	2.1	1.33

The comparative analysis of Tables 4.3 and 4.6 showed that the dispersive component determined by the vOCG method has greater values (except for the sample # 11) than that determined by the Berger method. For example, the dispersive component of the sample # 2 determined by the Berger method is 39.45 mJ/m^2, whereas that determined by the vOCG method is 43.85 mJ/m^2. On average, the difference constitutes 10–15%. According to the Berger method, the dispersive component can be found from the plot, by the Y-intercept, which is undoubtedly more accurate because ten liquids are used for plotting. Nevertheless, the results obtained by both methods are in qualitative agreement. Acidic samples estimated by the Berger method remain the same also in case of using nonlinear vOCG systems. Samples # 1, 2, and 6 are characterized by noticeable acidic parameters of 1.5, 2.23, and 2.48 mJ/m^2 correspondingly. The sample # 11, showing slight basic properties, is characterized in this case by non-zero basic parameter of 2.1 mJ/m^2.

A comparison of the acid-base SFE components demonstrated that the values obtained by different method substantially differ. This fact seems as not unexpected, since theoretical formulae for calculating γ_s^{ab} in both methods are different. The acid-base component determined by the Berger method has different value because of quite different calculation method using ten test liquids and graphical way of calculation.

Thus, a nonlinear modification of the vOCG method, despite some disagreements with the Berger method, provides consistent characteristics of acidic and basic surface properties of polyepoxides. Taking the drawbacks mentioned into account, the comparative analysis performed showed that both methods are suitable for qualitative estimation of the acidity or basicity of any solid surface. At the same time, the acidity parameter determined by the Berger method is rather easy in practical application. Therefore, a prediction of the behavior of polyepoxide-based adhesive jointed with metal, that is, the estimation of possibility to take part in acid-base interaction with substrate, will be more illustrative if using the aforementioned approach. For the moment, the Berger method using the acidity parameter which characterizes both acidic and basic properties of the same coating seems as more preferred.

So, the estimation of acid-base surface properties of polyepoxides performed by both methods demonstrates the possibility to affect these properties by choosing curing agent.

4.1.3 MODIFICATION OF POLYEPOXIDES

In spite of all valuable properties, epoxy resins become brittle enough when cured. To eliminate this drawback, different nonreactive as well as reactive modifiers can be introduced into formulation, being incorporated into the network structure and making the cured composites highly impact resistant, wear proof, flexible, etc. In this concern, one of the most promising is PEF-3A grade low molecular weight epoxy urethane rubber.

Polyurethane-based epoxy coatings are flexible and wear proof, highly impact- and crack resistant.

Introducing reactive modifiers into composition makes the curing of epoxy amine compositions more complicated. The new concurrent chemical curing reactions will occur, namely the reactions of functional groups of modifiers with primary and secondary amino groups. In this case, system investigation of the effect of reactive modifiers on the type of chemical reactions and curing process rules as well as on physical-mechanical and adhesive properties of epoxy amine systems is needed.

The types of chemical reactions which occur upon introducing PEF-3A rubber and the rules of epoxy amine system curing have been studied in detail [161]. Modified compositions show greater conversion rates, with the times of structural transitions decreasing due to flexible PTMG-based fragments of modifier. The question is whether surface acidity varies as a function of PEF-3A content in composition.

As the rubber content in composition increases, the acidity parameter of surface happens to decrease to a minimum at 70% PEF-3A and 30% ED-20 content and, then, raises a little bit (Fig. 4.1). Such dependence may be explained by variation in the content of functional groups, which can act as acid-base centers on the surface of composition, as ED-20 and PEF-3A content in the composition is varied at the corresponding amine content.

As mentioned above, these groups include urethane groups (increasing in content), OH-groups (decreasing in content), $>\ddot{N}$-groups (decreasing in content), $-\ddot{O}$-groups (increasing in content), and aromatic rings (decreasing in content). At 100% epoxy oligomer content, the OH-groups and $>\ddot{N}$ -groups prevail, with the surface being mainly acidic due to greater mobility of the OH-groups (the acidity parameter $D=5.1(mJ/m^2)^{1/2}$).

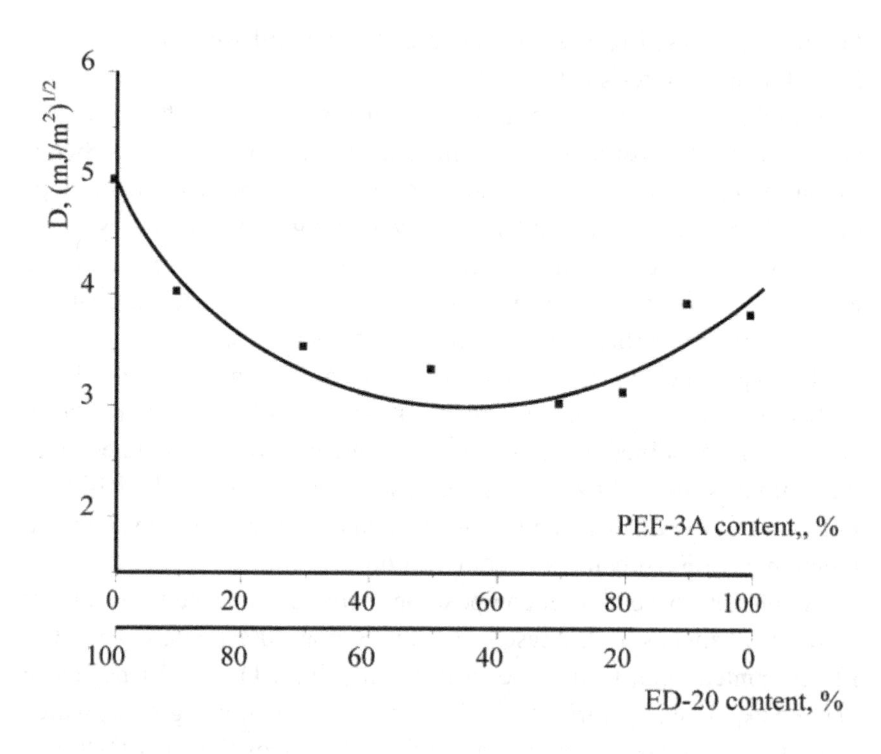

FIGURE 4.1 Acidity parameter as a function of ED-20 and PEF-3A content in composition.

Due to introducing PEF-3A rubber, the content of the OH-groups decreases and more -Ö-groups incorporated in PEF-3A rubber appear. This affects surface properties, and the acidity of surface becomes less pronounced. On the other hand, urethane groups start to act as acidic centers, and the acidity rises again upon curing compositions containing greater amount of PEF-3A rubber. As far as aromatic rings are concerned, their contribution to surface acidity appears to be insignificant due to steric hindrance [152, 154, 155].

Here, we should return to explanation of the dependencies shown in Fig. 2.12 (Chapter 2, Section 2.3.1) concerning the work of adhesion of formamide and glycerol. In view of above considerations regarding the changes in surface chemistry due to varying ED-20 and PEF-3A content in composition, it now becomes clear that the acidic glycerol and the basic formamide which are equal in the SFE value show different W_a values because of acid-base interaction on surface. Decrease in the acidity due to decrease in ED-20 content (Fig. 4.1) occurs symbiotically, as the W_a of the basic formamide which worse wets less acidic surface decreases (Fig. 2.12). At 100% rubber content and the increased D value, the W_a of formamide again raises. Similarly, the dependence for glycerol shown in Fig. 2.12 may be explained.

To return to modification by ED-20 resin, it should be noted that by varying PEF-3A content the targeted variation of acid-base properties of composition can be performed to achieve the optimal acidity parameter value.

4.1.4 EFFECT OF FILLERS AND PIGMENTS

Talc and chromium oxide being usually introduced in equal proportions are widely used as fillers and pigments in epoxy amine compositions. Pigment surface patterns, including Cr_2O_3, as well as talc surface properties were estimated [148] to determine their compatibility with film-forming agents. The effect of these additives on acid-base properties of coating is shown in Fig. 4.2.

The composition containing 54% ED-20 epoxy resin and 14% PEF-3A epoxy urethane rubber was used. As talc content rises (chromium oxide content decreases), the acidity of surface increases. In this case, great amounts of pigment and filler (up to 30%) exceed their critical content in composition. Presumably, the dependence observed may be explained by greater acidity of talc ($3MgO \cdot 4SiO_2 \cdot H_2O$) in comparison with chromium oxide the surface of which is neutral.

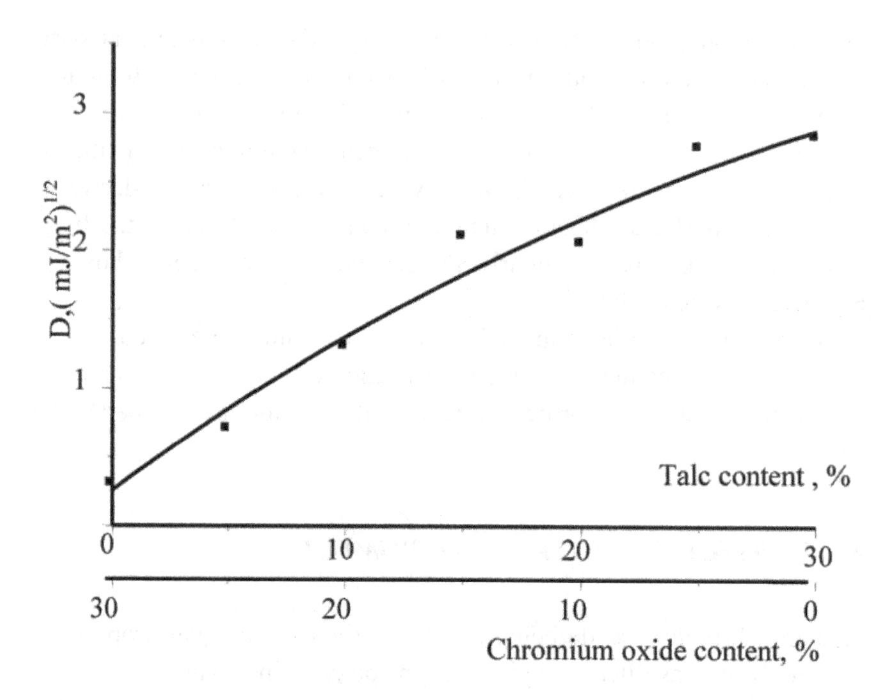

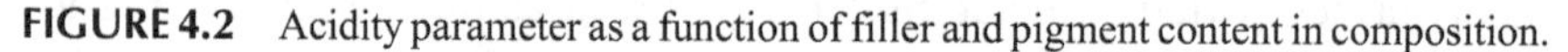

FIGURE 4.2 Acidity parameter as a function of filler and pigment content in composition.

4.2 ACID-BASE PROPERTIES OF POLYOLEFIN COMPOSITIONS

We shall consider the possibility of acid-base modification of polyolefins. Polyolefins used as protective coatings on metals are low- and high density polyethylenes (LDPE and HDPE) and ethylene-vinyl acetate copolymer (EVA) with variable content of vinyl acetate groups (VA-groups), modified by different additives. Ethylene-ethyl acrylate copolymer (EEA), ethylene-butyl acrylate copolymer (EBA), and ethylene-vinyl acetate-maleic anhydride terpolymer (EVA terpolymer), as well as terpolymers based on ethylene, vinyl acetate, and maleic anhydride in combination with different EVA copolymers produced by radical polymerization under elevated pressure were examined as promising adhesives.

4.2.1 EFFECT OF FORMATION CONDITIONS

Table 4.7 summarizes data on the acidic properties of unmodified polyethylene coatings produced by different methods. Thus, the sample # 1 was prepared by melting a fine powder (315 μm particles) at 200°C during 10 min, samples # 2 and 3 – at 220°C during 20 min, and samples # 4 and 5 – by press molding at 180°C during 5 min.

The difference in surface acidity of different grades of unmodified polyethylenes is negligible. The acidity parameters range within 2–4 $(mJ/m^2)^{1/2}$. This is the evidence of a low acidity of polyethylene surface due to thermal oxidation products evolved during formation process. As both the temperature (T_f) and time (t_f) of coating formation increase, the increase in the D value is observed (2.35 for the sample # 1 and 3.59 for the sample # 2) which may be well explained by the increased degree of polyethylene oxidation in this case. This goes to prove that the acidity parameter is very susceptible even to a slight variation of surface status. As exemplified by samples # 4 and 5, the exposure to air over a month results in no variation in properties in question.

TABLE 4.7 Acidity parameters of polyethylene coatings.

No.	Surface	D, $(mJ/m^2)^{1/2}$
1	HDPE weakly oxidized, T_f=200°C, t_f=10 min	2.35
2	HDPE strongly oxidized, T_f=220°C, t_f=20 min	3.59
3	LDPE grade 16803–070, T_f=220°C, t_f=20 min	4.05
4	LDPE grade 15303–003, T_f=180°C, t_f=5 min	3.1
5	LDPE grade 15303–003, T_f=180°C, t_f=5 min over a month	3.1

Thus, the choice of process conditions of coating formation can insignificantly affect acid-base properties of coatings.

4.2.2 MODIFICATION OF POLYETHYLENE COATINGS

Modification of polyolefins leads to variation in surface properties of polymer compositions. This equally refers to polyethylene- and EVA-based systems. Table 4.8 shows data on acid-base properties of coatings based on different grades of polyethylene modified by multifunctional additives. Acid-base properties were estimated according to the Berger acidity parameter.

TABLE 4.8 Acidity parameters of modified polyethylene coatings.

No	Surface	D, $(mJ/m^2)^{1/2}$
1	HDPE + 0.5% DH	2.21
2	HDPE + 1% DH	1.98
3	HDPE + 2% DH	0.62
4	HDPE + 5% DH	-1.28
5	LDPE + 0.25% MPDM	6.55
6	LDPE + 0.5% MPDM	6.45
7	LDPE + 1% MPDM	7.96
8	LDPE + 1.5% MPDM	6.52
9	LDPE + 4% MPDM	7.08
10	HDPE + 0.5% OPD	-0.50
11	HDPE + 0.5% pyrocatechol	7.05
12	HDPE + 2% DPP	7.61

Analysis of the results showed that modifiers containing primary amino group facilitate decreasing the acidity parameter of polyethylene coating contact surface down to negative values. In other words, the coatings modified by primary aromatic amines (DH modifier, OPD) are pronouncedly basic. This is correct, since primary amines are Lewis bases.

On the contrary, phenolic compounds being Lewis acids facilitate increasing the acidity parameter of coatings [108]. Increase in the acidity in the presence of m-phenylenedimaleimide is due to electrophilic fragment with double bonds. The last one, thanks to conjugation of nitrogen atoms with the phenyl ring as well as an acceptor effect of transannular carbonyl fragments in the maleimide ring, is a pronounced Lewis acid showing the minimal electron density of the –CH=CH– fragment known to be a strong electron acceptor able to form π-complexes with metals. IR-studies showed that not less than 20% of unreacted unsaturated maleimide fragments are still present in the coating formed [162].

Even the lowest modifier concentrations (0.25–0.5%) can profoundly alter the D value. This goes to prove that the acidity parameter is a characteristic susceptible even to slightest variation in composition formulation.

4.2.3 MODIFICATION OF ETHYLENE-VINYL ACETATE COPOLYMER COATINGS

Ethylene-vinyl acetate copolymers are widely used as adhesives in multilayer anticorrosion coatings. Since EVA layer must show excellent adhesive ability to outer layer as well as to metals and primers, examination and modification of surface characteristics of this layer within the framework of the acid-base approach is of great interest.

The acidity of unmodified EVA varies as the content of VA-groups increases, which follows from the dependence visualized in Fig. 4.3.

This is correct, since the VA group incorporates the carbonyl group showing the basic properties due to greater electronegativity of the oxygen atom against the carbon atom. The basicity of surface rises as the content of VA groups increases up to 20% followed by the increase in the D value, probably because of steric hindrance due to the excess VA groups traveling onto the surface.

The variation in surface properties is also observed in case of polyisocyanate (PI) modified EVA (Fig. 4.4).

PI-modified EVA shows more pronounced surface basicity, since PI being introduced as promoter of adhesion, first, contains itself basic active functional groups and, second, the hydration of modifier by atmospheric moisture can yield carbamic acids which easily lose CO_2 to form Lewis bases, namely amines and ureas [150, 151].

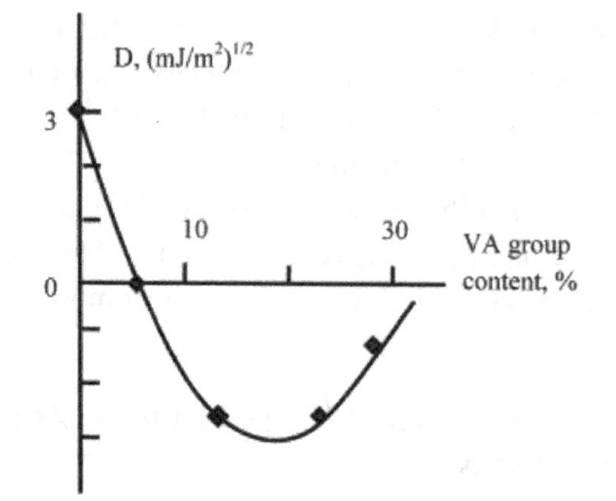

FIGURE 4.3 Acidity parameter of EVA-based coating as a function of VA group content.

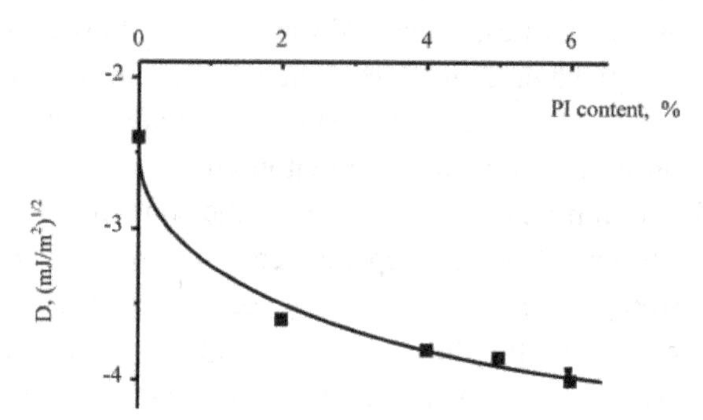

FIGURE 4.4 Acidity parameter of EVA-based coating as a function of PI content.

Again, the second dependence demonstrates that slight variations in composition formulation affect acid-base properties of composition.

4.2.4 USING THE VOCG METHOD FOR ETHYLENE-VINYL ACETATE COPOLYMER COATINGS

The vOCG method for non-linear systems yields interesting results concerning acidic and basic properties of coatings based on SEVA-14 grade EVA modified by ethyl silicate (ES) in the presence of catalyst of various percentage. Modification of polymers by silicone compounds profoundly affects the crystallinity, the nature of interchain bonds, packing density in amorphous regions of crystallizable polymers, and, hence, the total set of physical-mechanical properties and rheological behavior. Major applications of silane modified polymers today are cables, polymer pipes for hot water, adhesives, glues, coatings, primers, foamed and heat shrink parts, packaging materials.

The results obtained are visualized in Figs. 4.5 and 4.6. As seen from the figures, the dispersive and total SFE values of coating are not virtually susceptible to variations in modifier content; only at 10% modifier both values do slightly decrease. It can be easily explained by the fact that excess modifier travels onto surface followed by the decrease in the SFE, probably because a modifier itself has less SFE compared with EVA. Variation in the acidic parameter occurs within the experimental error (Fig. 4.6, Curve # 1), whereas the basic parameter shows maximal value at the concentration of 5–7%.

Chemical interaction of ES with EVA may be assumed to be effective only if the ES content is not greater than 5%, and surface basicity increases probably due to traveling of basic siloxane groups onto surface.

As modifier content continues to grow, the modifier probably starts to act as a filler to form three-dimensional network in polymer matrix and segregate into separate phase.

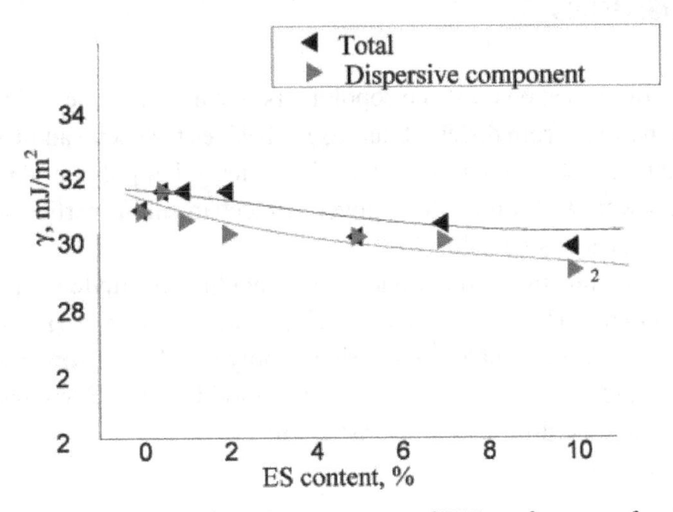

FIGURE 4.5 SFE and its dispersive component of EVA surfaces as a function of ES content.

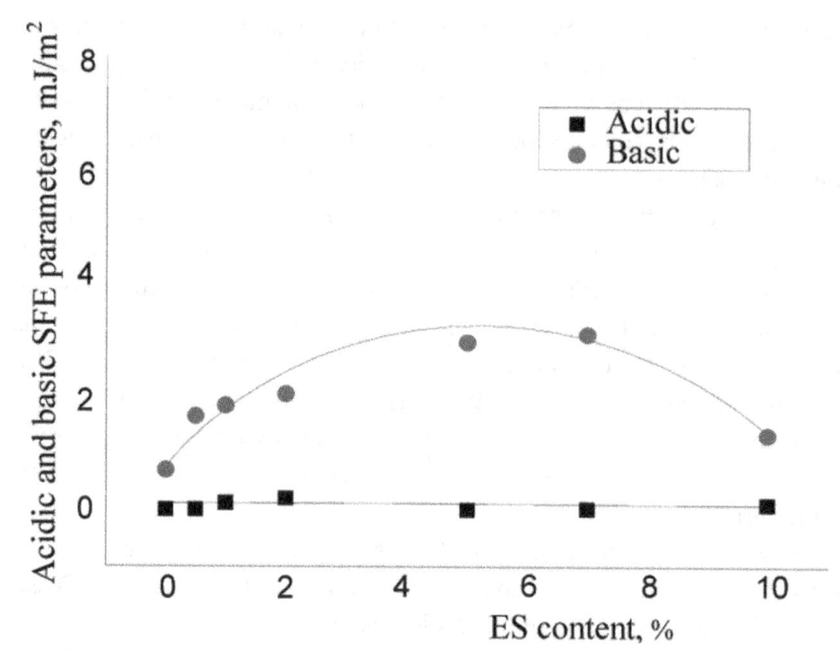

FIGURE 4.6 Acidic and basic parameters of EVA surfaces as a function of ES content.

4.3 ACID-BASE PROPERTIES OF ETHYLENE COPOLYMER COMPOSITIONS

Compositions based on ethylene copolymers find a fairly wide and diverse application today, from different purpose adhesives to special additives for inks. Whatever the application field, it is required that polymer compositions show well wettability, great physical-mechanical properties, and, of course, a high adhesive ability.

For estimating the surface energy and acidity of ethylene-ethyl acrylate copolymer (EEA), ethylene-butyl acrylate copolymer (EBA), and ethylene-vinyl acetate-maleic anhydride terpolymer (EVA terpolymer) the Berger method was used. The surfaces were modified by ES and tested.

Energy and acid-base characteristics of initial and modified samples are presented in Table 4.9.

TABLE 4.9 SFE components and acidity parameters of ethylene copolymer compositions.

Sample	γ_s (mJ/m²)	γ_s^d (mJ/m²)	γ_s^{ab} (mJ/m²)	D, (mJ/m²)$^{1/2}$
EEA				
unmodified	34.5	28.9	5.6	1.8
ES 3%	37.9	34.0	3.9	1.7
ES 5%	34.9	30.7	4.2	0.95
ES 7%	38.45	29.7	8.75	0.95
ES 10%	34.0	28.3	5.7	1.1
EBA				
unmodified	33.8	28.1	5.7	3.0
ES 3%	35.5	30.9	4.6	2.5
ES 5%	35.3	29.4	5.9	1.95
ES 7%	33.3	28.3	6.0	2.3
ES 10%	32.9	28.9	4.0	2.2
EVA terpolymer				
unmodified	41.4	27.9	13.5	4.1
ES 3%	37.0	27.3	9.7	3.4
ES 5%	37.7	24.4	13.3	3.2
ES 7%	40.1	27.1	13.0	3.8

The results were obtained by using no catalyst. Introduction of ES into EEA results in slightly decreased surface acidity, which achieves its minimum at ES content of 5–7%. On the contrary, introducing ES in greater amounts facilitates the increase in the acidity parameter, with no regularity of variation in both acid-base and dispersive SFE components of samples being observed. The same may be said in case of using both EBA and EVA terpolymer.

However, a considerable effect of introducing silicone compounds into polyolefins (silane cross-linking) is known to be observed only in the presence of catalysts among which organotin compounds are preferred. Catalyst initiates a chemical interaction of ES with copolymer, which results in the incorporation of the siloxane unit into a side group [163]. Therefore, the examination of samples modified by ES in the presence of catalyst is of special interest (Table 4.10). According to the results, the tendency of the acidity of films to decrease is observed, as the ES content increases. This fact can be rationalized, since a chemical reaction of silane modification can yield in this case ethyl acetate known to be basic [163, 164].

TABLE 4.10 SFE components and acidity parameters of ethylene copolymers modified by ES in the presence of catalyst.

Sample	γ_s (mJ/m²)	γ_s^d (mJ/m²)	γ_s^{ab} (mJ/m²)	D, (mJ/m²)$^{1/2}$
EEA				
unmodified	34.5	28.9	5.6	1.8
ES 5%	34.9	28.9	6.0	1.4
ES 10%	34.7	28.8	5.9	0.4
EBA				
unmodified	33.8	28.1	5.7	3.0
ES 3%	35.1	28.4	6.7	2.75
ES 5%	35.3	28.8	7.5	2.2
ES 7%	35.7	29.0	6.7	1.3

In addition to compositions mentioned, the examination of acid-base properties and surface energy were performed using a fairly complicated systems, namely ethylene-vinyl acetate-maleic anhydride terpolymers (OREVAC terpolymers) in combination with different ethylene-vinyl acetate copolymers (EVATANE copolymers) produced by high pressure radical polymerization. For some of these composition materials, a super-additive increase in adhesive properties is observed at a certain ratio of components [165].

The results obtained by the Berger method for compositions based on SEMA 9307 and SEMA 9305 grade EVA terpolymers in combination with SEVA 20-20 and SEVA 28-05 grade EVA copolymers are shown in Table 4.11. All samples contain 10% of talc.

In each of three cases, a little variation in composite surface characteristics, the acidity in particular, can be seen. A reasonable question is what the cause of this effect is. Obviously, no chemical interactions occur in the mixture of these compounds, and all depends probably on structural rearrangements taking place in composite as the ratio of components is varied.

A noticeable acidity of virtually all surfaces except unmodified SEVA-20 grade EVA is observed. It was mentioned earlier that, when in a great amount, the VA groups are probably too sterically hindered to travel onto surface, which may lead to positive D values of the tested systems [166]. As the terpolymer content decreases, the acidity gradually goes down. We state the fact that the variation in component percentage in polymer mixtures also affects both acidity and basicity of polymer surfaces.

TABLE 4.11 SFE components and acidity parameters of ethylene copolymer mixtures.

Sample	γ_S (mJ/m^2)	γ_S^d (mJ/m^2)	γ_S^{ab} (mJ/m^2)	D, (mJ/m^2)$^{1/2}$
Series 1 SEMA 9307 + SEVA 20				
SEMA 9307–100%	40.95	29.2	11.75	3.2
SEMA 9307–50% SEVA 20–50%	37.6	27.9	9.7	2.85
SEMA9307–30% SEVA 20–70%	39.1	26.8	12.3	3.3

TABLE 4.11 *(Continued)*

Sample	γ_s (mJ/m²)	γ_s^d (mJ/m²)	γ_s^{ab} (mJ/m²)	D, (mJ/m²)$^{1/2}$
SEVA 20–100%	36.7	29.5	7.2	0.6
Series 2 SEMA 9305 + SEVA 20				
SEMA 9305–100%	37.4	28.4	9.0	4.2
SEMA 9305–60% SEVA 20–40%	34.9	27.7	7.2	3.95
SEMA 9305–50% SEVA 20–50%	35.6	28.3	7.3	3.85
SEMA 9305–20% SEVA 20–80%	39.0	26.7	12.3	4.3
SEVA 20–100%	36.7	29.5	7.2	0.6
Series 3 SEMA 9305 + SEVA 28				
SEMA 9305–100%	37.6	28.1	9.5	4.2
SEMA 9305–60% SEVA 28–40%	38.4	27.6	10.8	3.75
SEMA 9305–30% SEVA 28–70%	39.9	27.9	12.0	4.0
SEVA 28–100%	35.2	26.6	8.6	2.45

4.4 ACID-BASE PROPERTIES OF MODIFIED RUBBERS

Among the existing rubber adhesives, of special interest are adhesives to be applied onto adhesive primer, intended for use in tape insulants for different purpose pipes. Besides advantages, these materials have a number of significant drawbacks, with one of them being a low strength of adhesive joint in aggressive environments or under drastic temperature difference. In this concern, a metal – rubber primer – rubber adhesive type system is of significant scientific interest.

Rubber base of compositions defines both a unique property set of coatings derived therefrom and the possibility to control major mechanical characteristics by choosing the required amounts of components in formulations. These systems are composite materials composed of a great number of components introducing each of them alters physical-mechanical properties of polymer.

Examination of surface characteristics of rubber systems may facilitate predicting and further enhancement of interaction at both metal – primer and primer – rubber adhesive type interfaces. In this concern, the problem was solved stepwise, that is, the examination was made of surface properties of model surfaces which were unmodified rubbers, modifiers themselves, and compositions based on rubbers with various percentages of modifiers [110, 166, 167]. Acidity parameters of unmodified rubbers were given in Table 3.4, section 3.9.4 of previous chapter.

Authors studied the effect of the nature of modifiers (adhesion promoters, cross-linking agents, and other multifunctional additives) on wetting of rubber primers of adhesive tapes by test liquids. A large series of chlorobutyl rubber coatings with variable content of different modifiers was examined. Thus, it was found that the acidity parameter of rubber composition surface decreases, as the content of manganese dioxide rises. The same effect was observed in case of increasing p-Quinone dioxime (PQD) content. As is known, PQD is a low temperature vulcanizing agent which, in some cases, improves the adhesion of chlorine-containing rubbers to steel [168].

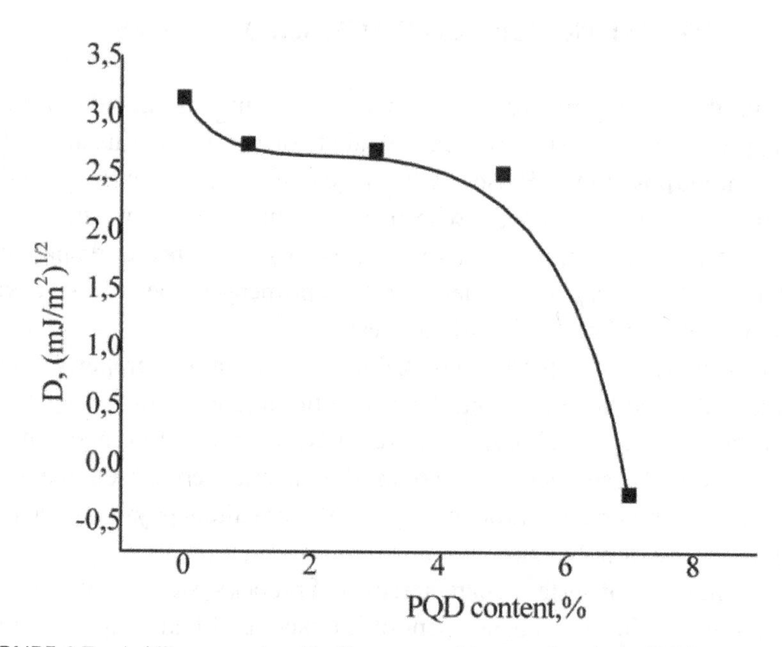

FIGURE 4.7 Acidity parameter of rubber composition as a function of PQD content.

It was revealed that introducing more than 5% of PQD results in the acidity parameter noticeably decreased down to negative values, in other words, in the increased surface basicity (Fig. 4.7). The concentration of 5% seems to be optimal. Further introduction of modifier leads to its traveling onto surface. It is interesting that adding PQD and increasing its content results in the increased force of peeling of the tape with both vulcanized adhesive and primer layers from steel, with the most pronounced increase occurring at modifier concentration of 5%. PQD is known to dissolve in many organic solvents and can be evenly dispersed in bulk of composition, which allows a great deal of bonds (physical and chemical) between polymer and steel surface [169]. Studies carried out by using the Berger method also confirm the variation of the nature of primer surface, which results in its acidity decreased down to negative values. Once the PQD content increases, the decrease in the acidity is natural, since this modifier shows pronounced properties of a Lewis base due to the nitrogen atom with a lone electron pair.

Modification of acid-base properties of compositions containing PQD as vulcanizing agent occurs under the influence of adhesion promoters, particularly, manganese dioxide MnO_2 which is a strong PQD oxidizer (Fig. 4.8). Introducing MnO_2 enhances the force of peeling of vulcanized composition from steel at both temperatures of 20°C and 80°C [169].

This might be caused by activation of oxidative processes upon coating formation. Oxidative processes seem to play a substantial role in the increase of adhesion the PQD-vulcanized chlorobutyl rubber shows to steel surface. Transition metal oxides are not known to cause fast rubber degradation. Therefore, introducing those into composition is more preferred than introducing, for example, cobalt stearate [170]. It follows from Figure 4.8 that the increase in the MnO_2 content results in the acidity parameter of surface decreased down to negative values (from 2.5 in the absence of MnO_2 down to -0.95 at 10% content thereof).

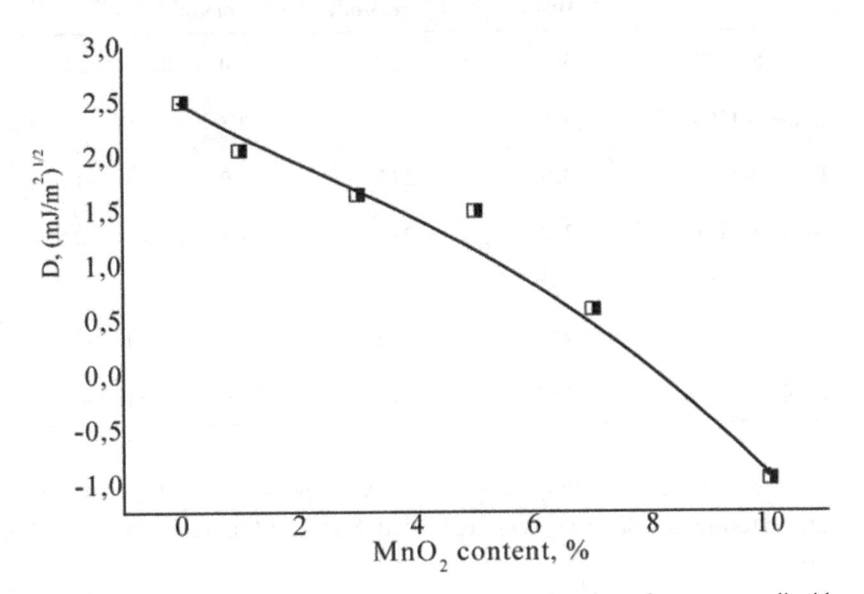

FIGURE 4.8 Acidity parameter of rubber primer as a function of manganese dioxide content.

The decreased acidity at the increased MnO_2 content is explained by the fact that it oxidizes PQD, contained in composition, to p-dinitrobenzene (PDB) which, in its turn, shows basic properties due to the nitroso group.

In recent years, different polymeric petroleum resins (PPR) are widely used as rubber modifiers. Eskorets series PPRs are high molecular weight polydiene type compounds derived by stepwise catalytic polymerization of the C_5 fraction. These contain a little portion of unsaturated compounds and are susceptible to oxidation, with the addition of atmospheric oxygen occurring via adjacent double bonds. Carbonyl groups released thereby facilitate decreasing the acidity parameter of PPR surface. Surface-energy and acid-base characteristics of model PPR film sample surfaces are presented in Table 4.12. As Eskorets-1401 grade PPR content increases, the acidity parameter of surface rises (Fig. 4.9).
****

TABLE 4.12 PPR surface characteristics.

PPR grade	γ_S (mJ/m^2)	γ_S^d (mJ/m^2)	γ_S^{ab} (mJ/m^2)	D, (mJ/m^2)$^{1/2}$
Eskorets-1102	30.2	26.2	4.0	2.1
Eskorets-1304	35.3	32	3.3	2.5
Eskorets-1310	32.6	27.7	4.9	–1.2
Eskorets-1401	31.2	29.4	1.8	–2.8
Eskorets-M5	36.7	29.4	7.3	–3.3
Piroplast-2	37.5	33.4	4.1	0.9
Himplast	37.5	30.2	7.4	–0.9

To explain the results, the proton NMR spectra of PPR solutions in carbon tetrachloride [171] were recorded. Some of them are shown in Fig. 4.10.

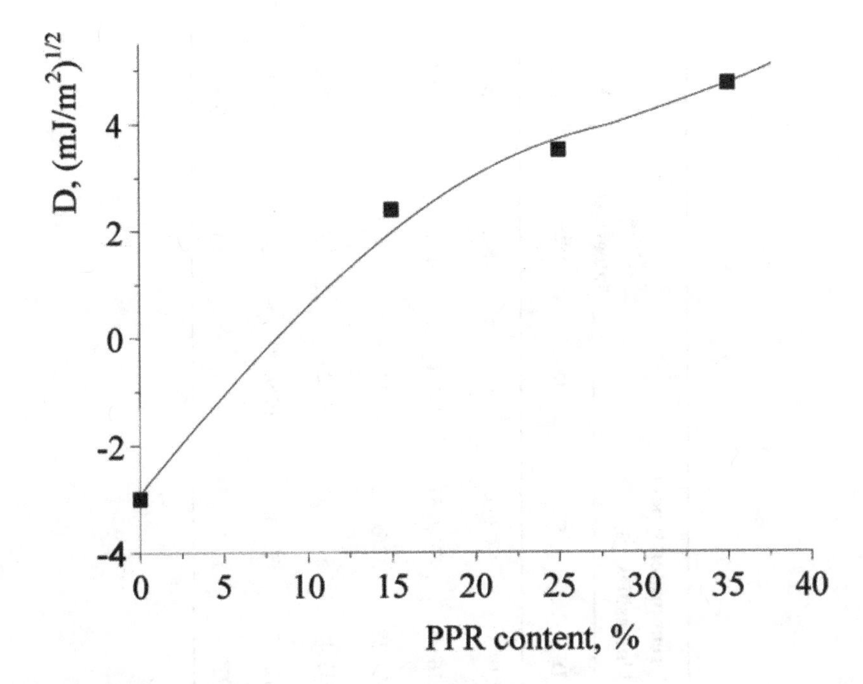

FIGURE 4.9 Acidity parameter of rubber primer as a function of Eskorets -1401 grade PPR content.

Protons in PPR spectra can be divided into 6 types which are characterized by the following chemical shifts [172]: A – aromatic (6.2–8.0 ppm), B – olefin (4.0–6.2 ppm), C – methyl α and methylene α to a benzene ring (1.0–3.6 ppm), D – methine of paraffins and naphthenes (1.5–2.0 ppm), E – methylene of paraffins and naphthenes (1.05–1.5 ppm), F – methyl (0.5–1.05 ppm). The relative content of protons of different functional groups was determined by using the integrated proton NMR spectrum. Integrated intensity values of the six proton types for each PPR sample as well as normalized integrated intensity values, that is, a percent ratio of the number of protons of different types obtained by multiplying the integrated intensity of proton groups of each PPR sample by the corresponding factor K, are presented in Table 4.13.

TABLE 4.13 PPR characteristics according to proton NMR data.

PPR grade	Integrated intensities of different proton types / Normalized integrated intensities, %						Total integrated value	K
	A	B	C	D	E	F		
Eskorets-1102	0.25/0.13	5.5/3.0	14/7.57	32/17.3	48/26.0	85/46.0	184.75	0.541
Eskorets-1304	1.5/0.85	4.67/2.6	15/8.5	30/16.93	47/26.51	79/14.6	177.2	0.564
Eskorets-1401	0/0	8.5/4.76	13/7.28	38/21.28	37/20.72	82/45.94	178.5	0.56
Eskorets-M5	2/1.2	0/0	14/8.83	63/37.48	51/30.35	38/22.62	168	0.595
Himplast	30/14.9	11/5.45	52/25.9	56/27.9	33/16.4	19/9.45	201	0.498
Piroplast-2	16/14.1	5.5/4.85	35/30.8	27/23.8	19/16.75	11/9.7	113.5	0.881

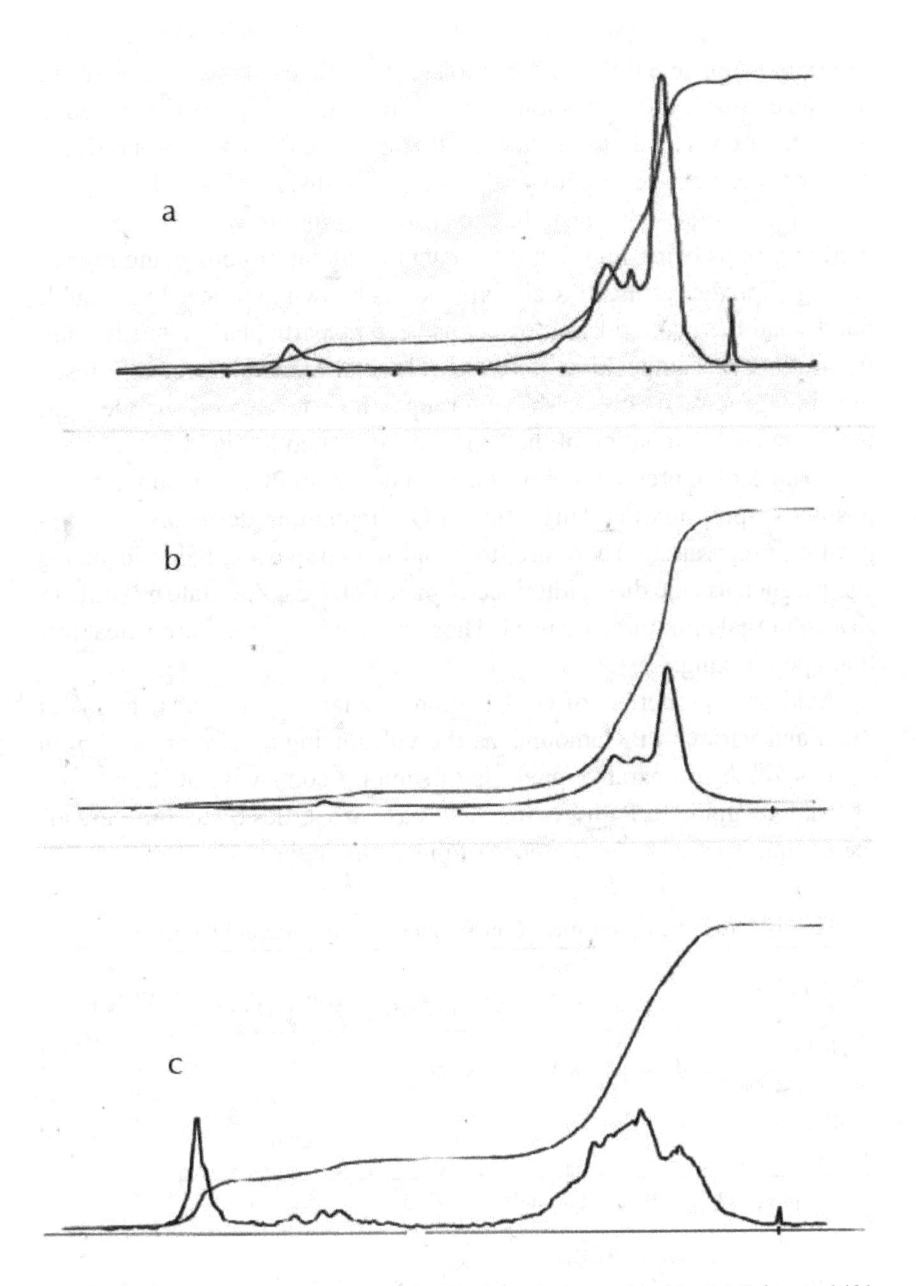

FIGURE 4.10 Proton NMR spectra of PPRs in carbon tetrachloride: (a) Eskorets-1401, (b) Eskorets-1102, (c) Himplast.

As seen from Fig. 4.10 and Tables 4.12 and 4.13, PPR samples differ by composition and the acidity parameter D. The results obtained indicate that the compounds in question, differing by acid-base properties (according to the Lewis acid-base theory), are able to affect surface properties of rubber adhesive in different ways, having been dispersed therein.

Thus, Eskorets-1401 resin has no aromatic protons, with the content of methyl groups being high but the content of olefin protons being highest among aliphatic resins. PPR air oxidation is known to proceed via double bonds, and thermal oxidation reactions become particularly intensive during milling of composition at 100°C. Thermal oxidation seems to result in acidic groups, such as carboxyl groups, to be formed on surface. This would be the explanation of the dependence plotted in Fig. 4.10.

A substantial premature vulcanization occurs in PQD-containing compositions upon mixing. This makes PQD inappropriate to primer composition preparation. Therefore, to avoid this drawback, the vulcanizing agents, such as zinc diethyldithiocarbamate (ZDEC), ZnO-talc mixture, or ZDC-ZnO-talc mixture, are used. Those induce less premature vulcanization upon mixing.

Acid-base properties of compositions containing 10% of talc, 1% of ZnO, and various PPR amounts as the vulcanizing system are shown in Table 4.14. A comparative analysis of samples containing 30% and 40% of Eskorets grade PPR shows that acid-base properties of coatings are affected in this case by the presence of this resin grade.

TABLE 4.14 Acid-base properties of compositions as a function of PPR content.

Sample[1]	γ_S (mJ/m^2)	γ_S^d (mJ/m^2)	γ_S^{ab} (mJ/m^2)	D, (mJ/m^2)$^{1/2}$
CIIR 59% Eskorets 30%	34.1	30.9	3.2	6.95
CIIR 49% Eskorets 40%	29.5	26.5	3.0	—0.45

[1]All samples contained 10% of talc and 1% of ZnO.

The acidity parameter of sample #1 is substantially greater (6.95 compared with that of sample #2 which is virtually zero), which indicates a high surface acidity. The increase of the resin content (up to 40%) leads to the decreased acidity parameter of –0.45, which may be explained by traveling of the excess amount of the resin onto surface, which makes it more basic (it is worthwhile noting that the acidity parameters of Eskorets-1310 and Eskorets-1401 based films are –1.2 and –2.8 correspondingly). The PPR content of 30% seems to be optimal. Thus, the modifier studied allows acid-base properties of rubber surfaces to be varied in rather broad ranges, with the acidity being primarily increased.

The adhesive properties of talc-ZnO based systems are not too high, so adhesion promoters for primer layer are required. The 2,2,4-trimethyl-1,2-dihydroquinoline oligomer (TDQ) (Russian technical TDQ grade is Acetonanyl-R) was suggested [169] which allows high values of the force of peeling of the vulcanized tape from steel. Acetonanyl-R based coating shows the acidity parameter of –0.85 and, hence, is rather basic. Acid-base properties of rubber primers modified by various amounts of this modifier are shown in Fig. 4.11. These results were obtained by using the Berger method.

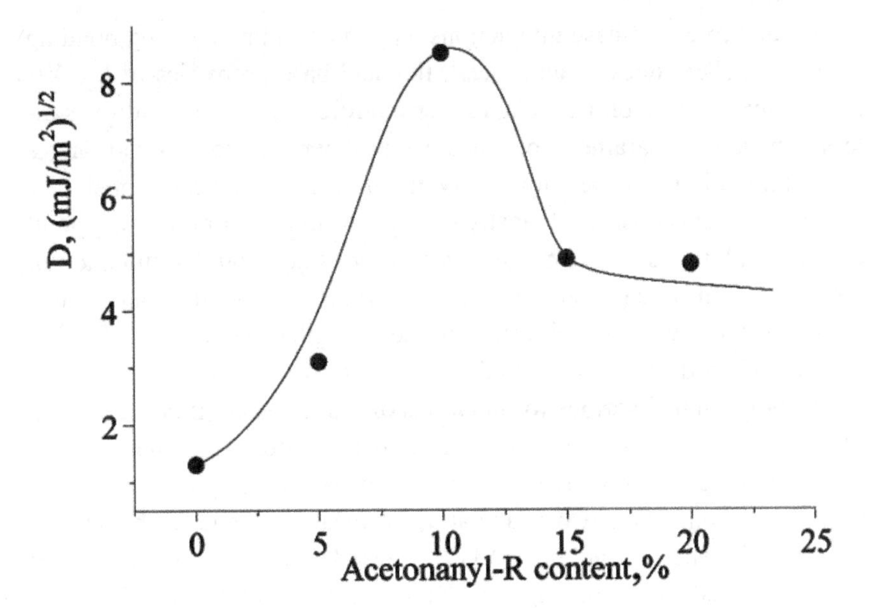

FIGURE 4.11 Acidity parameter of rubber primer as a function of Acetonanyl-R grade TDQ content.

However, introduction of Acetonanyl-R into rubber in the absence of PPR results in no increase in adhesive force of primer-steel joint. This is explained by heterogeneity of Acetonanyl-R – rubber mixture, with Acetonanyl-R itself having a low adhesion to chlorobutyl rubber. A high value of the force of peeling upon simultaneous introducing PPR and Acetonanyl-R allows the assumption of their synergistic effect on the adhesion of the vulcanized chlorobutyl rubber. The reasons of this effect will be analyzed further in view of the acid-base approach.

Ultimately, the determination of surface characteristics of rubber compositions according to the Berger method makes the examination of the effect of any given modifier on surface properties of composition possible, as well as allows optimal concentrations in formulations to be found for further forecasting and targeted controlling interaction at metal-primer and primer – rubber adhesive interfaces.

4.5 ACID-BASE PROPERTIES OF BREAKER RUBBER MIXES

To reveal a role acid-base interactions play in providing a strong bonding between rubber adhesive and metal, the acid-base properties of breaker rubber mixes based on isoprene rubber modified by diverse additives altering the acidity parameter of vulcanized rubber surface in broad ranges were studied in the same way. Today, the problem of adhesion in similar systems is being considered mostly from the formulation-processing point of view, making no reckoning of acid-base consideration. Examination of the role of acid-base interactions played in vulcanizate – brass-plated steel cord systems as well as their effect on the strength of vulcanizate – brass-plated steel cord contact is, therefore, of particular scientific interest.

Authors studied various formulations of model rubber mixes based on SKI-3 grade isoprene rubber and vulcanizing system (polymeric sulfur + Sulfonamide M), modified by diverse cobalt-containing adhesive additives (Monobond 680C (Co-B-acylate), cobalt stearate, and Luch 10 grade cobalt naphthenate) (Table 4.15) [173–178]. The prepared rubber mixes were then used for producing the plates in contact with brass similar in composition to brass used for plating of steel cord for the tyre industry.

These plates were produced by press molding under various process conditions (155°C, during 15 and 25 min).

TABLE 4.15 Composition of model rubber mixes (wt. parts).

No	Component	Formulation					
		I	II	III	IV	V	VI
1	SKI-3 isoprene rubber	100	100	100	100	100	100
2	P-245 carbon black	—	29	58	58	58	58
3	Polymeric sulfur	7.5	7.5	7.5	7.5	7.5	7.5
4	Sulfonamide M	1.0	1.0	1.0	1.0	1.0	1.0
5	Monobond 680C (Co-B-acylate)	—	—	—	0.4	—	—
6	Luch 10 cobalt naphthenate	—	—	—	—	1.0	—
7	Cobalt stearate	—	—	—	—	—	1.0

Surface-energy characteristics and the acidity parameter values of rubber coatings are presented in Table 4.16. It follows therefrom that both rubber mix composition and temperature-time conditions of vulcanization substantially affect surface characteristics and the acidity of samples. Thus, coatings based on similar formulation but varied in preparation time show rather different acid-base properties, with the absolute difference between the acidity parameter values being 5.2 $mJ^{1/2}/m$ (sample # II). The acid-base SFE component of this sample varies twofold. The total SFE of coatings in all cases varies slightly, max 9% (for samples #IV and V).

TABLE 4.16 Surface characteristics of rubber mixes vulcanized in combination with brass.

Sample	g^d, (mJ/m²)	g^{ab}, (mJ/m²)	g, (mJ/m²)	D, (mJ/m²)$^{1/2}$
		155°C, 15 min		
I	26.8	9.9	36.7	–0.25
II	26.4	7.25	33.65	–0.1
III	23.1	14.6	37.7	1.2
IV	26.1	6.7	32.8	–4.2
V	27.2	7.9	35.1	–1.4
		155°C, 25 min		
I	25.4	11.4	36.8	1.5
II	31.5	3.2	34.7	5.2
III	30.4	6.8	37.2	2.6
IV	23.8	11.2	35.0	–3.25
V	29.0	3.2	32.2	3.0
VI	31.4	6.3	37.7	–2.6

4.6 ACID-BASE PROPERTIES OF METAL SUBSTRATES

As the information of acid-base characteristics of metal is needed for producing coatings with optimal adhesive properties, different metal surfaces were examined in this concern [149, 150]. Copper, titan, brass, and different grades of steel and duraluminum were used as metal supports. The acidity parameters measured are presented in Table 4.17. The results obtained demonstrate a rather broad range of the acidity parameter values which vary from –2.2 (mJ/m²)$^{1/2}$ for D16 grade duraluminum to 5.5–8.1 (mJ/m²)$^{1/2}$ for different steel grades. Most values are positive, which indicates a predominant acidity of metal surfaces. Usually, metal substrates are Lewis acids.

TABLE 4.17 Surface characteristics of metals.

Metal	g_s^d, mJ/m^2	g_s^{ab}, mJ/m^2	γ_s, mJ/m^2	D, (mJ/m^2)$^{1/2}$
Titan	25.1	15.4	40.5	0.9
Copper	24.8	7.2	32.0	3.3
L62 brass	23.1	13.3	36.4	4.0
L90 brass	26.1	14.0	40.1	8.4
Ya1T steel	24.2	11.3	35.5	2.9
St3 steel	23.2	10.6	33.8	1.7
St10 steel	26.3	8.3	34.6	8.1
St20 steel	22.0	19.3	39.3	5.5
G65 steel	24.5	14.3	38.8	4.55
EI696 steel	22.2	19.2	41.4	1.3
ChZh1 black sheet metal	18.5	19.2	37.7	6.5
Aluminum	26.7	15.0	41.7	–1.9
D16 duraluminum	28.5	15.6	44.1	–2.2
D16T duraluminum	25.5	21.1	46.6	4.2
D16AM duraluminum	21.0	24.1	45.1	3.2
D16ATB duraluminum	25.2	19.1	44.3	2.7

Metals and metal oxides are known to be referred to high-energy surfaces showing the SFE of about 500 mJ/m^2. The question is how those relatively not high values may be explained. Properties of metal surfaces are substantially affected by the environment, and this is particularly observable in case of oxide-covered metals. High-energy surfaces adsorb atmospheric water vapors and other admixtures, such as organic substances, which lead to the lowered SFE of substrate [3, 179].

The studies performed [109] showed that thermal treatment of metal has the varying effect on the D value. In case of copper, aluminum, and zinc-plated sheet metal, the exposure of metal to 150°C during 10–15 min leads to the increased acidity parameter, whereas for titan, brass, and steel such a treatment results in the decreased D value. This effect seems to be caused by the fact that oxides, which are formed on metal surface upon thermal treatment, are of different nature, which is defined by the chemistry of surface itself. As is known for iron [180], Fe_3O_4 and Fe_2O_3 oxides are acidic, whereas surfaces of substrates like the hydrated Fe_2O_3, hydroxides, or FeO are basic.

In case of St3 grade steel, the effect of prolonged air oxidation at room temperature during two months on surface-energy and acid-base properties was studied in detail. The St3 acidity parameter just after the treatment (grinding and degreasing with acetone) is 1.25 $(mJ/m^2)^{1/2}$, the next day – 1.7 $(mJ/m^2)^{1/2}$, gradually rising over the following two weeks up to 3.3 $(mJ/m^2)^{1/2}$. Further, over a month and two months, the D value remains unchanged. Figures for all substrates listed in Table 4.17 refer to the surface status observed not later than a day from the moment of treatment.

The Berger method implies a substantial effect of solvent used as degreaser on the acidity parameter of substrates [101]. Thus, the acidity parameters of the same surface differ in case of acetone and CH_2C1_2. As such, this fact is interesting and can find practical application, but for the sake of uniformity of comparative estimation, acetone was used in all experiments carried out by authors.

It is worthwhile noting that different steel grades profoundly differ by acid-base properties of surface. The grades St3, St10, and St20 refer to mild and medium steels, with the carbon content being 0.03, 0.1, and 0.2% correspondingly. These do not show any noticeable corrosion resistance, and a major contribution to their surface acidity is due to the oxide film formed. To verify this statement, the examination of St3 and St20 surfaces was performed by scanning electron microscopy (SEM) using the X-ray electron microprobe analysis. Elemental composition of surfaces is shown in Table 4.18. Microphotographs and secondary radiation spectra of surface elements are visualized in Figs. 4.12 and 4.13.

TABLE 4.18 Elemental composition of steel surfaces.

Element	St3 grade steel		St20 grade steel	
	wt.%	atom %	wt.%	atom %
C	5.69	20.86	4.43	17.73
Si	0.92	1.44	0.14	0.23
Fe	90.97	71.78	95.43	82.04
O	2.04	5.61	—	—
Mn	0.39	0.31	—	—
Total	100	100	100	100

According to electron microprobe analysis data, the St20 surface does not incorporate oxides, and, once metal atoms are Lewis acids, this probably explains a high acidity of this steel grade (D = 5.5 $(mJ/m^2)^{1/2}$).

On the contrary, the presence of oxygen on St3 surface proves that iron oxides present on metal surface are predominantly basic, which results in the D value decreased down to 1.7 $(mJ/m^2)^{1/2}$. Moreover, a few amounts of manganese and silicon were detected. These elements are introduced into similar steel grades during production (Mn 0.3–0.7%, Si 0.2–0.4%) [181].

Steel grades Ya1T and EI696 refer to nickel-chromium alloy high steels. A combination of chromium and nickel improves physical and chemical properties of steel. Chromium is known to increase corrosive resistance of steel under atmospheric conditions and forms Cr_2O_3 surface oxides which are neutral. This is probably a reason that acidity parameter values of these materials are lower than those of St20 grade steel.

Microphotographs in Fig. 4.12 and 4.13 point to the roughness of substrate surfaces, which is due to production conditions and grinding during measurement preparations. Treatment of all metal surfaces was carried out in the same way; therefore the roughness coefficient of all

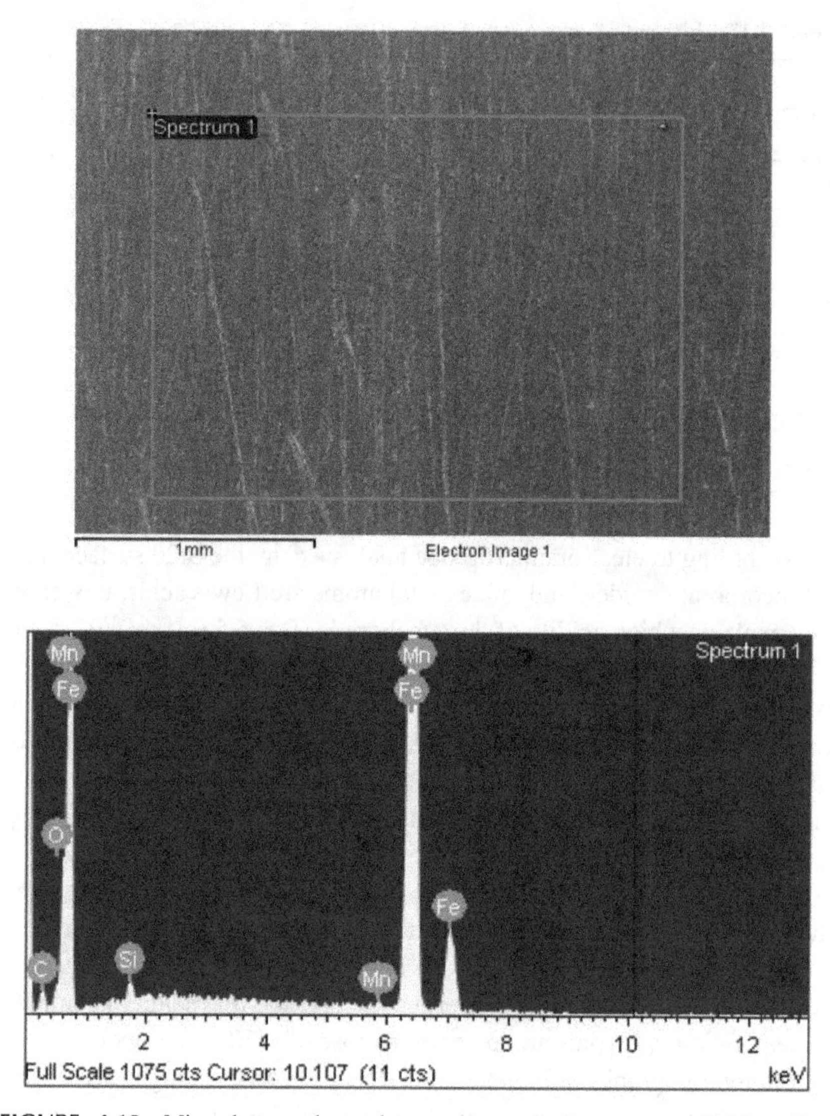

FIGURE 4.12 Microphotographs and secondary radiation spectra of St3 surface elements.

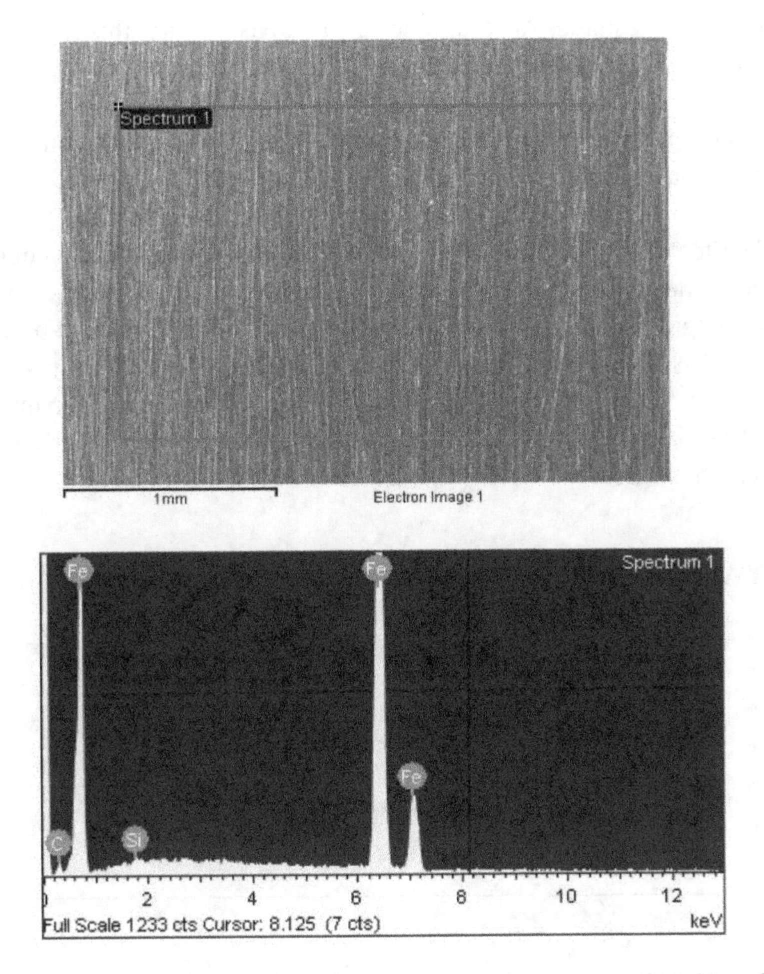

FIGURE 4.13 Microphotographs and secondary radiation spectra of St 20 surface elements.

Substrates is approximately the same and was not taken into account in determination of the values presented in Table 4.17.

The material outlined in the present chapter proves conclusively that there is a broad range of challenges to affect acid-base characteristics of polymer surfaces, namely, by varying process conditions for coating formation, introducing specific modifiers, varying component ratio, etc. To solve adhesive problems of each particular coating system, an individual approach is required. According to the acid-base theory, this implies the choice of adhesives and adherents, which would show maximal given difference in acid-base characteristics (the acidity parameter). It is the acidity and basicity of polar molecules incorporated into composition of condensed phases that are crucial factors in interaction of a liquid with another liquid, solids with liquids, and polymers with solid substrates. The material systematized in this chapter is reference information, since it contains data on the acidity and basicity of polymers commonly used in adhesive technologies as well as modifiers used in compositions. The acid-base approach has been recently recognized by the leading international researchers in the field of adhesion. The Chapter 5 shows that acid-base interactions are largely responsible for the formation of adhesive bonds acting across the interface.

KEYWORDS

- **Berger method**
- **metal substrates**
- **polyepoxide**
- **polyolefin**
- **rubber**
- **vOCG method**

CHAPTER 5

ACID-BASE APPROACH TO THE PROBLEM OF INTERACTION ENHANCEMENT IN ADHESIVE JOINTS

CONTENTS

The acid-base approach to the problem of interaction enhancement in adhesive joints is a promising theory which, due to its versatility, could facilitate solving many practical adhesion problems. According to this approach, the best adhesive joint is achieved provided that one of joinable materials shows acidic properties, while the other one has basic ones. Therefore, the main task in designing adhesive joint is to achieve maximal difference in those properties of adhesive and adherend, although getting such a difference is necessary but not always sufficient condition for excellent adhesive interaction because of other operating adhesion mechanisms.

Acid-base properties of adhesives and adherents, that is, polymeric composition materials and metal substrates, have been considered in detail in Chapter 4. The present chapter will concern the application of the acid-base theory to the explanation of adhesive patterns observable in testing the strength of adhesive joints based on the materials studied.

5.1 THE ROLE OF ACID-BASE PROPERTIES PLAYED IN ADHESION OF POLYMER COATINGS OF DIFFERENT NATURE

In many cases, acid-base interaction of adhesive with substrate is a crucial factor in the formation of adhesive bonds acting across the interface. The greater the probability of the formation of acid-base bonds between joinable materials, the higher the strength of adhesive joint, which was confirmed by numerous publications. A direct relationship between acid-base characteristics of adhesives and adherents and the strength of adhesive joints prepared is observed in many papers [101, 182–185]. Finlayson et al. state that acid-base interaction is a crucial factor which defines bond strength in the ethylene-acrylic acid copolymer – aluminum system, since the adhesion strength rises as the content of acidic groups of adhesive and/or basic groups of substrate increases [182]. Targeted preparation of surface, such as glass treatment by silane-containing agents or polyvinyl acetate [77], can substantially alter and even reverse the acid-base nature for providing necessary adhesive bonds. Berger [101] also verified the correlation between the shear strength of lap joints and the acidity of samples and pointed out a noticeable effect of substrate rigidity on shear strength values. The best correlation is achieved in case of more flexible substrates.

The fact of the leading role of acid-base interactions in adhesion was verified by using many metal-polymer and polymer-polymer systems. Leggat et al. report that the increasing adhesion in silicone-epoxy film joints can be explained by acid-base interaction only [183]. The treatment of epoxy coatings by silanes [184] results in the increased adhesive interaction with aluminum below the isoelectric point, which is also possible due to acid-base interaction. Luner points out a substantial contribution of cellulose esters, which are basic on bipolar substrates showing various SFE values to the work of adhesion [140].

For enhancing adhesion of carbon and glass fibers to epoxy groups, the acid-base approach was used upon surface treatment of components [186–187]. McCafferty experimentally verified that the acid-base approach is true for enhancing adhesion of polymers to oxide-covered metals [188]. Thus, the force of peeling of the acidic adhesive increases provided that metal surface is covered by the basic oxide. At the same time, the force of normal peeling of the basic polymer, such as polymethyl methacrylate, also increases provided that metal surface is covered by the acidic oxide.

The relationship between adhesive properties and acid-base properties was verified by authors of this monograph by a series of examinations of polymeric coatings, such as polyolefin, rubber, and epoxy ones most widely used in adhesive technologies [107,109–111,150–154,158, 160, 165–166, 174–176, 189–211].

5.2 THE RELATIONSHIP BETWEEN ACID-BASE AND ADHESIVE PROPERTIES OF POLYEPOXIDE COATINGS

In view of the aforesaid, of special interest is the verification of possibility to estimate the adhesive ability of material and to predict the efficiency of adhesive interaction in a joint, using the obtained (and reported in the previous chapter) acid-base surface characteristics. The leading role of the difference in acid-base properties of adhesive and adherend has been emphasized earlier. Therefore, using the term "relative acidity parameter" would be reasonable for correct estimating the adhesive ability:

$$DD = |D_{coating} - D_{substrate}| \tag{5.1}$$

The growth of the relative acidity parameter indicates the increase of the difference in acid-base properties of polymer and substrate and should be accompanied by the increase in acid-base interaction, with other conditions being equal. As the DD is an absolute difference of parameters, it is obvious that adhesive may act against support as either an acid or a base. If a compound has a great amount of diverse functional groups able to adhesive bonding, this variety of functional groups will allow interactions with both acidic and basic substrate. As the acidity or basicity of substrate increases against the adhesive, intermolecular interactions enhance, and the joint becomes more strengthened.

Adhesive interaction of all epoxy coatings and substrates was estimated by the resistance to cathodic disbandment [212]. Samples including different curing agents were tested. The results of testing three coating groups using curing agents of various activities are presented in Fig. 5.1. PEPA, AF-2M grade aminophenol, and DTB-2 grade product were used as highly, medium, and low active curing agents correspondingly.

The experiment performed allowed the assumptions suggested earlier to be verified, namely: each of three groups shows that the rate of interaction between surfaces increases as the difference in their acid-base properties (the ΔD) raises, which results in the diminished defect diameter.

For the coatings based on PEF-3A-modified ED-20 grade epoxy resin, their resistance to cathodic disbandment from St3 grade steel was also studied. The results obtained are shown in Fig. 5.2.

A comparison of data resulted from Figs. 4.1 and 5.2 showed that compositions containing not greater than 70% of PEF-3A demonstrate a correlation between the adhesive interaction in joint and surface acidity. Again, this results to conclusion that the greater the difference in acid-base properties of surfaces, the less defect diameter of coating. As the acidity of adhesive against the substrate increases, intermolecular interactions enhance, and the joint becomes more strengthened. The increase in rubber content up to 70%, as follows from Fig. 4.1, leads to the decreased D value of composition, therefore defect diameter rises. However, at greater PEF-3A contents, that is, the surface acidity goes up again, the resistance to cathodic disbandment would be expected to enhance according to the present approach. What can be actually observed is the decrease in adhesive interaction. As stated by Berger [101], one of the difficulties in interpreting these results is that a good correlation between the acidity parameter and adhesive interaction force is observed if the rigidity of adhesive is comparable to that of adherend. In this case, the predominance of rubber in composition leads to alteration of physical-chemical properties and rheological behavior of material, which affect the formation of a contact with support. A certain deterioration of adhesive properties of coating is observed in this case.

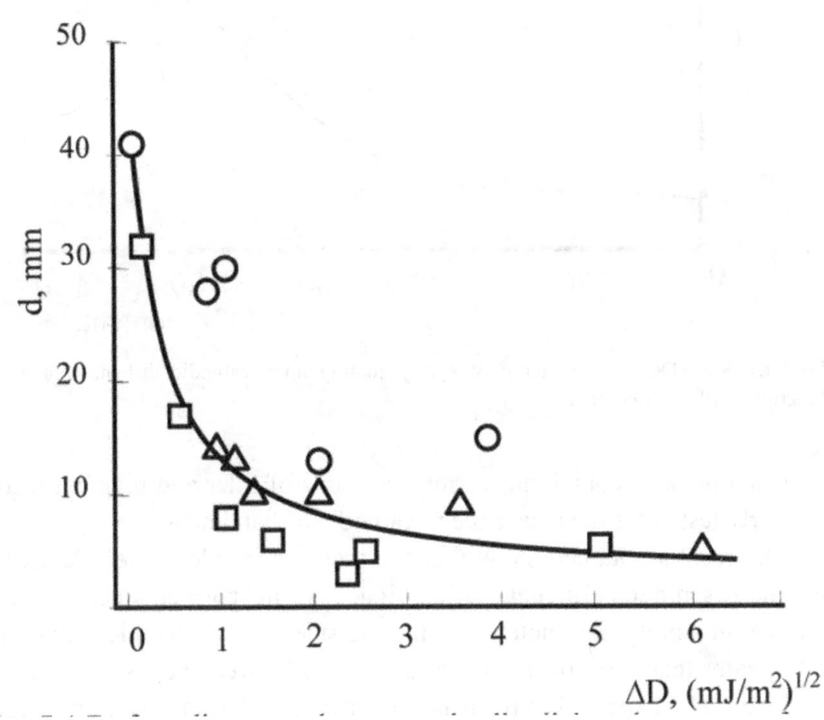

FIGURE 5.1 Defect diameter d upon cathodic disbandment as a function of the ΔD of metal – epoxy coating adhesive pairs (○ – ED-20 + PEPA; □ – ED-20 + AF-2M; Δ – ED-20 + DTB-2).

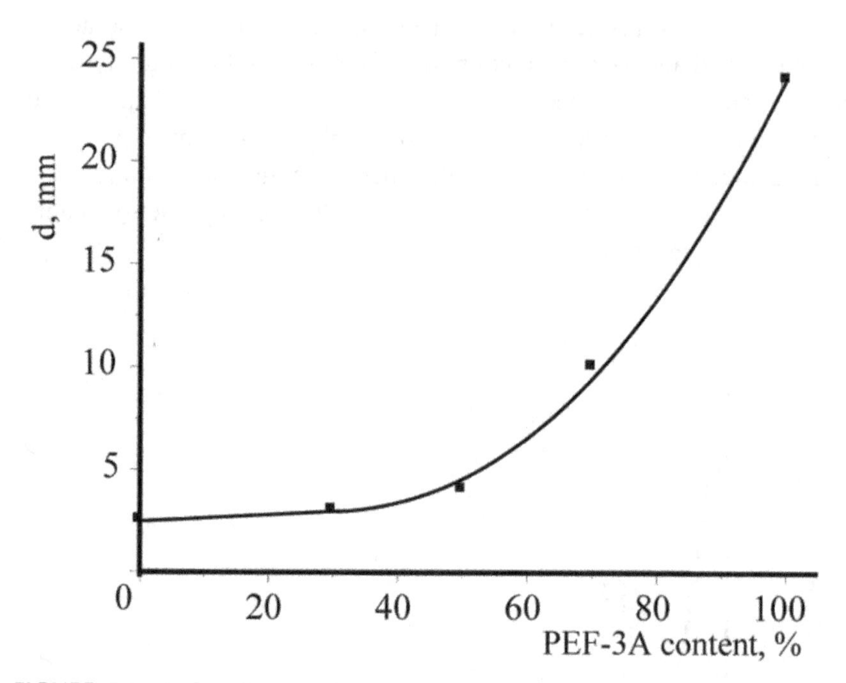

FIGURE 5.2 Defect diameter d of epoxy coating upon cathodic disbandment as a function of PEF-3A content.

Compositions containing various amounts of filler and pigment were similarly tested for the resistance to cathodic disbandment (Fig. 5.3).

The results obtained show that the increase in talc content leads to the increased defect diameter of coating. It seems correct, since the difference in acidity parameters against the steel for the samples containing greater than 15% of talc is negligible, and a weak adhesive interaction would be expected in this case according to the acid-base approach (Fig. 4.2, Chapter 4). In this experiment, coatings were prepared on the oxidized steel surface (D = 3.3 $(mJ/m^2)^{1/2}$), and it is easy to verify that the resistance to cathodic disbandment decreases as the ΔD value approaches to zero.

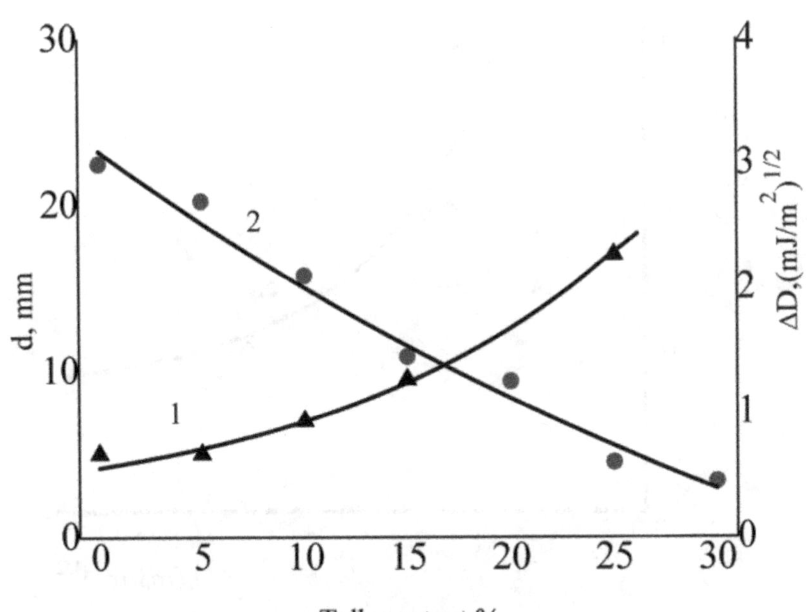

FIGURE 5.3 Defect diameter d (Curve 1) and the ΔD value (Curve 2) as a function of filler content.

Cathodic disbandment of both epoxy compositions cured by CS # 1–9 curing systems (Table 4.4, Chapter 4) and coatings based on commercial epoxy powder paints (EUROKOTE-714.41, EX-4413F102, PE-507191, Scotchcote-226N grades) formed on St3 grade steel under routine process modes, was performed. Again, the obtained results presented in Figs. 5.4 and 5.5 verify that assumptions of the correlation between acid-base and strength properties of adhesive joints are true. The resistance to cathodic disbandment of all polyepoxides studied is defined by the difference in the acidity of interacting surfaces and increases as this difference goes up.

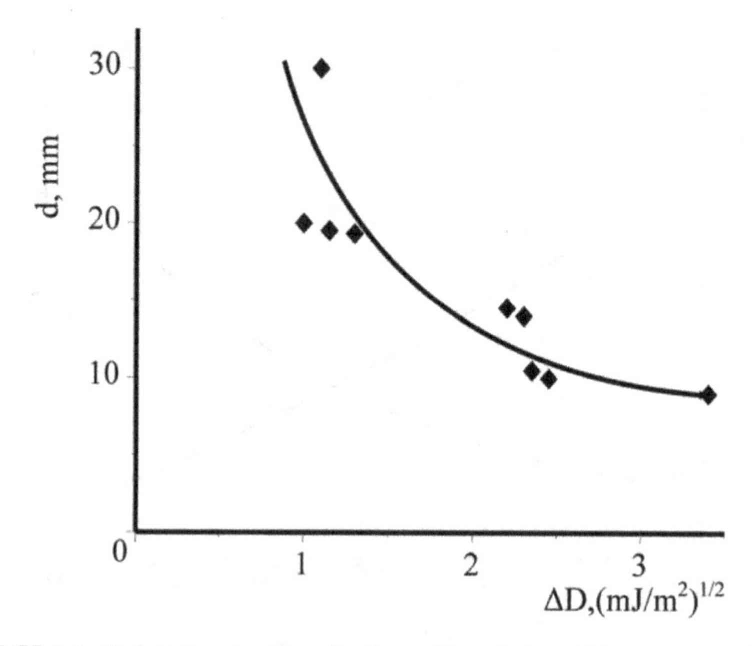

FIGURE 5.4 Defect diameter d as a function of the relative acidity parameter of ED-20+CS – St3 adhesive pairs.

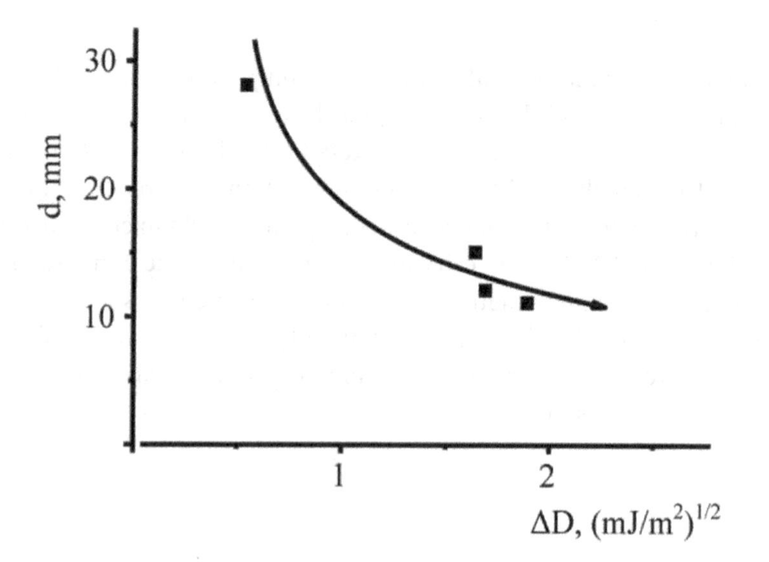

FIGURE 5.5 Defect diameter d as a function of the relative acidity parameter of epoxy powder paints on St3.

The resistance to cathodic disbandment of epoxy coatings cured by curing systems # 1–12 based on Lewis acids and tri(halogenalkyl) phosphates (Table 4.5, Chapter 4) from St3 grade steel is demonstrated in Fig. 5.6. Polyepoxides using the same curing agent were prepared on steel supports of various oxidation degree (D=1.25 and 3.3 $(mJ/m^2)^{1/2}$) to create more ΔD values.

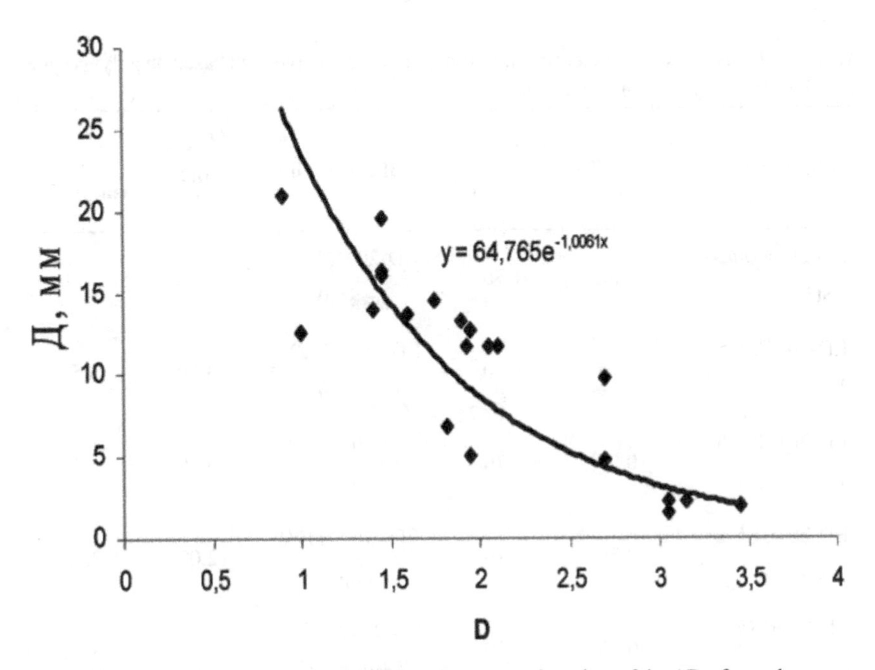

FIGURE 5.6 Resistance to cathodic disbandment as a function of the ΔD of metal – epoxy coating pairs using curing agents based on Lewis acids and tri(halogenalkyl) phosphates:

Figure 5.6 shows that the decrease in defect diameter is accompanied by the increase in the difference in acid-base properties of joinable materials. The dependence observed is near exponential and can be approximated with the exponent visualized in the figure as a solid line. The results obtained correlate with the data obtained earlier for ED-20/PEF-3A-based coatings using different amine type curing agents.

Experimental data for all metal – epoxy coating systems using amine type curing agents are summarized in Table 5.1, where the variety of metal substrates can be seen. The plot summarizing all dependencies obtained is visualized in Fig. 5.7, where all data groups for polyepoxides (Figs. 5.1–5.6) are contained. Such a generalization became allowed because similar curve patterns were observed for all groups.

Thus, the relationship between defect diameter (d) of all polyepoxide coatings studied upon cathodic disbandment and the relative acidity parameter ΔD of adhesive systems mentioned (Fig. 5.7) has been found.

TABLE 5.1 Acidity parameters of the components of adhesive joint based on polyepoxides cured by amine type curing systems.

Adhesive joint	$D_{polymer}$, $(mJ/m^2)^{1/2}$	D_{metal}, $(mJ/m^2)^{1/2}$	Adhesive joint	$D_{polymer}$ $(mJ/m^2)^{1/2}$	D_{metal}, $(mJ/m^2)^{1/2}$
ED-20+Kr*oo*t-1 +St3	–2.30	1.70	ED-20+DTB-2 + St20	2.00	5.50
ED-20+CS# 5 + St3	–1.65	1.70	ED-20+AF-2M + St20	3.00	5.50
ED-20+CS# 6 + St3	–0.95	1.70	ED-20+PEPA + St20	5.10	5.50
ED-20+CS# 1 + St3	–0.70	1.70	ED-20+DTB-2 + copper	2.00	3.30
ED-20+CS# 2 + St3	–0.60	1.70	ED-20+AF-2M + copper	3.00	3.30
ED-20+CS# 3 + St3	–0.50	1.70	ED-20+DTB-2 +L62	2.00	4.00
ED-20+CS# 4 + St3	0.35	1.70	ED-20+AF-2M +L62	3.00	4.00
ED-20+CS# 7 + St3	0.60	1.70	ED-20+PEPA +L62	5.10	4.00
ED-20+CS# 8 + St3	2.60	1.70	ED-20+DTB-2 + titan	2.00	0.90

TABLE 5.1 *(Continued)*

Adhesive joint	$D_{polymer}$, $(mJ/m^2)^{1/2}$	D_{metal}, $(mJ/m^2)^{1/2}$	Adhesive joint	$D_{polymer}$ $(mJ/m^2)^{1/2}$	D_{metal}, $(mJ/m^2)^{1/2}$
ED-20+CS# 9 + St3	2.85	1.70	ED-20+AF-2M + titan	3.00	0.90
Eurokote-714.41 + St3	0.10	1.70	ED-20+PEPA +D16T	5.10	4.20
EX4413-F102 + St3	1.20	1.70	ED-20+Af-2M +D16AM	3.00	3.20
PE50–7191 + St3	3.30	1.70	ED-20+PEPA +D16AM	5.10	3.20
Scotchcote226N + St3	3.55	1.70	ED-20+DTB-2 +Ya1T	2.00	2.90
ED-20+UP-583 + St3	6.80	1.70	ED-20+AF-2M + D16ATB	3.00	2.60
ED-20+DTB-2 + St10	2.00	8.10	ED-20+AF-2M +EI-696	3.00	1.30
ED-20+AF-2M + St10	3.00	8.10	ED-20+PEPA +EI-696	5.10	1.30

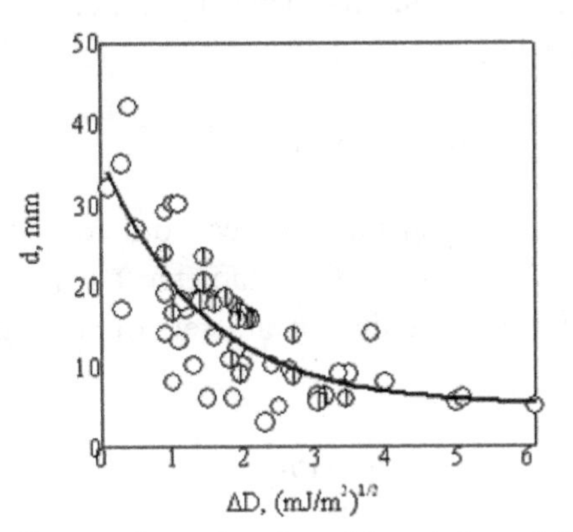

FIGURE 5.7 Defect diameter d as a function of the relative acidity parameter of metal – epoxy coating adhesive pairs: (⦶ – coatings based on ED-20 and tri(halogenalkyl) phosphates, ○ – coatings based on ED-20 and amine type curing agents).

The ΔD growth, that is, the increase of the difference in acid-base properties of polymer and substrate, is accompanied by enhancement of acid-base interaction estimated by defect diameter of coating under cathodic polarization.

The ΔD ranging from 0 to 2.7 $(mJ/m^2)^{1/2}$ indicates that interacting surfaces slightly differ in the acidity parameter. According to the acid-base approach, close ΔD values prove that functional groups of interacting surfaces are, predominantly, of similar nature, either acidic or basic, and can not therefore form acid-base bonds, which ultimately results in a weak adhesive interaction. A low resistance to cathodic disbandment in this region (an average defect diameter here exceeds 15 mm) is also the evidence. As the status of surface depends on the variety of factors (such as, air moisture, processing method, oxidation degree of metals), the presence, absence or variation of some factors can cause a slight deviation of the acidity parameter towards less or greater values and, hence, to the ΔD variation.

The effect of exterior conditions on both polymer and metal surfaces is not the same, which stipulates a large scattering of points in the plot.

The region of a strong acid-base interaction commences from $\Delta D > 2.7$ $(mJ/m^2)^{1/2}$, since defect diameter of coating in this region does not exceed 10 mm on the average. As the difference in acidity parameters rises, the interaction in adhesive joint enhances, but the scattering of experimental points decreases.

Experimental data are approximated by exponential curve, which is given by the expression:

$$d = d_0 + A\exp(-\beta\Delta D) \tag{5.2}$$

where d_0 has the meaning of initial defect formed in coating prior to the experiment (~5 mm). In fact, defect diameter tends to zero, the initial d_0 value, as the difference in acidity parameters of coating and substrate increases, i.e.:

$$d_0 = \lim_{\Delta D \to \infty} d \tag{5.3}$$

At $\Delta D=0$ the exponent becomes equal to 1, and $d = d_0 + A$. Hence, the A value has the meaning of maximal observed defect diameter of coating (in this case, $A \approx 40$ mm at fixed experimental time). At $\Delta D=0$ the acid-base interaction is minimal, and, hence, the resistance to cathodic disbandment is minimal, but defect diameter is maximal. The β parameter is reciprocal of such ΔD at which the transition from weak to strong interaction region takes place. In fact, $d = d_0+A/2.72$ at $\beta = 1/\Delta D$, and defect diameter is e=2.72 times less than maximum possible one, that is, $d \approx 15–20$ mm. The $\Delta D \approx 2$ satisfies these d values. Since almost all epoxy primers studied and the majority of metals are predominantly acidic, the attempt to achieve the ΔD values exceeding 6.1 $(mJ/m^2)^{1/2}$ has failed, which makes us find new curing agents for ED-20 grade epoxy resin and opportunities to modify metal surfaces as well.

The best adhesive ability estimated by the resistance to cathodic disbandment is observed for the following adhesive joints:

(ED-20+PEPA) – titan (ΔD = 4.6);

(ED-20+AF-2M) – St10 steel (ΔD = 5.1);

(ED-20+DTB-2) – St10 steel (ΔD = 6.1).

Figure 5.8 depicts possible model of acid-base interaction at the interface for one of strongest adhesive joints, namely (ED-20+AF-2M) – St10 steel (ΔD = 5.1).

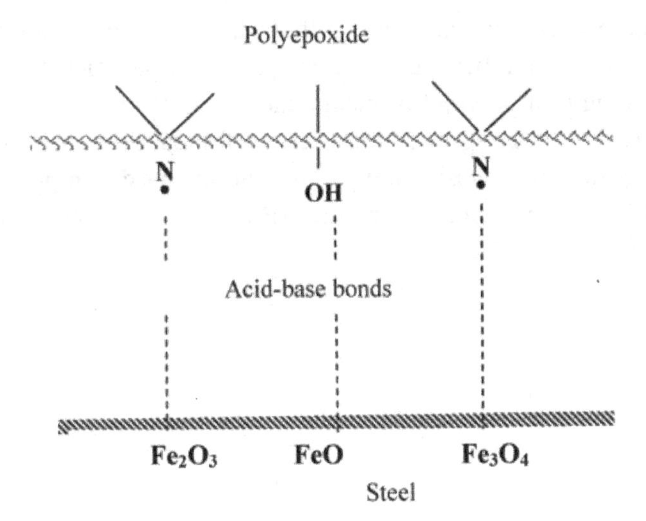

FIGURE 5.8 Model of possible acid-base interaction of polyepoxide adhesive with steel.

Examination of surface properties of metal substrates showed that St10 grade steel is acidic (D = 8.1 $(mJ/m^2)^{1/2}$), which may be explained only by the increased content of acidic (according to Lewis) iron oxides on surface. These are Fe_2O_3 and Fe_3O_4 oxides. This, however, does not rule out the fact of a total absence of basic oxides, such as FeO. The presence of both types of oxides in various percentages is the most probable. As far as adhesive is concerned, it was said earlier that the –OH and -groups act as functional on its surface. Hence, the assumption makes sense that a strong adhesive interaction observed is mainly due to donor-acceptor bonds formed between the nitrogen having a lone electron pair and acidic iron oxides. The interaction of hydroxyl groups of adhesive with basic iron oxides cannot be ruled out, but according to acidity parameters obtained, such an interaction occurs to a lesser extent. These considerations were exemplified by the model shown in Fig. 5.8.

Of course, the model does not cover all possible interactions in adhesive joint, with the most probable ones being visualized in the figure.

5.3 THE RELATIONSHIP BETWEEN ACID-BASE AND ADHESIVE PROPERTIES OF POLYOLEFIN COATING – METAL SYSTEM

Due to numerous examinations, authors have got the opportunity to verify the correlation between the relative acidity parameter of surfaces and the strength of polyolefin-metal adhesive joint. In particular, it was proved that the acidity growth for a series of metal substrates correlates with the growth of adhesive ability of coatings based on polyethylene modified by primary aromatic amines (PAAs) to those substrates (Fig. 5.9).

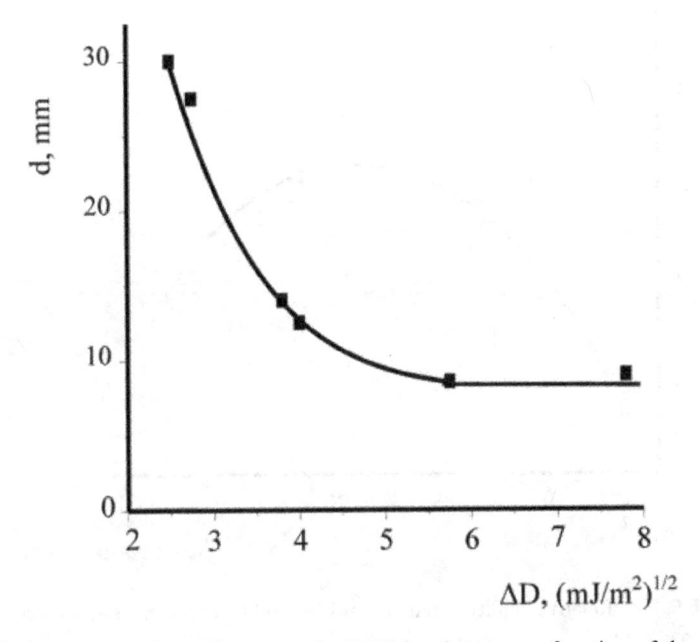

FIGURE 5.9 Defect diameter upon cathodic disbandment as a function of the relative acidity parameter of adhesive pairs: metal – polyethylene coating modified by DH modifier (2%).

Adhesive interaction was estimated by using the results of disbandment occurred under cathodic polarization. Steel, aluminum grades, black sheet metal, titan, copper, and brass were used as metal substrates. According to the results obtained, the greater the D of metal substrate, the less minimal defect diameter of the corresponding adhesive joint.

Besides polyethylene adhesives, the adhesive interaction of EVA copolymers incorporating VA groups of various percentages with St3 steel surface was similarly studied. In case of EVA used as adhesive, enhanced adhesive properties are also required against both polyolefins and metals or epoxy primers.

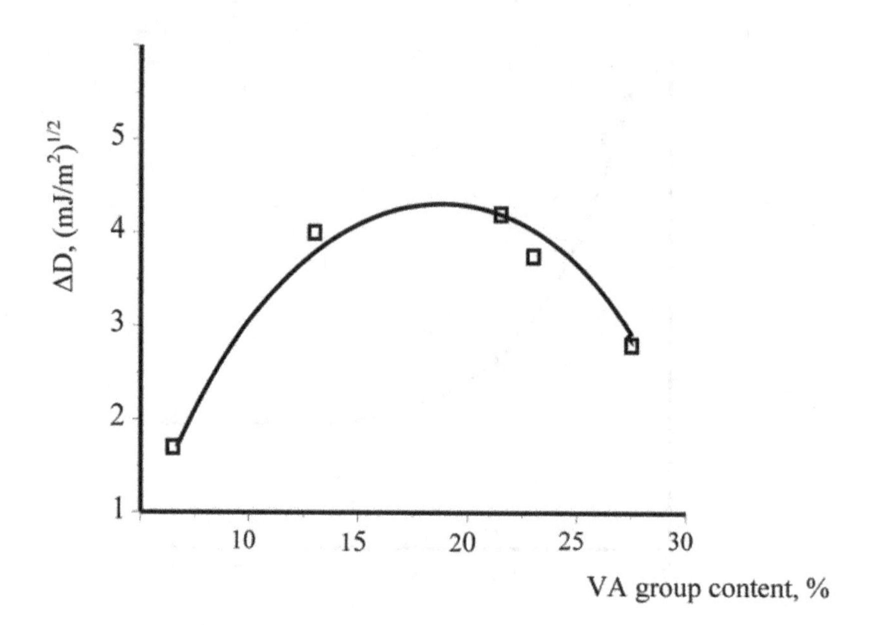

FIGURE 5.10 Relative acidity parameter of EVA-St3 adhesive pairs as a function of the VA group content.

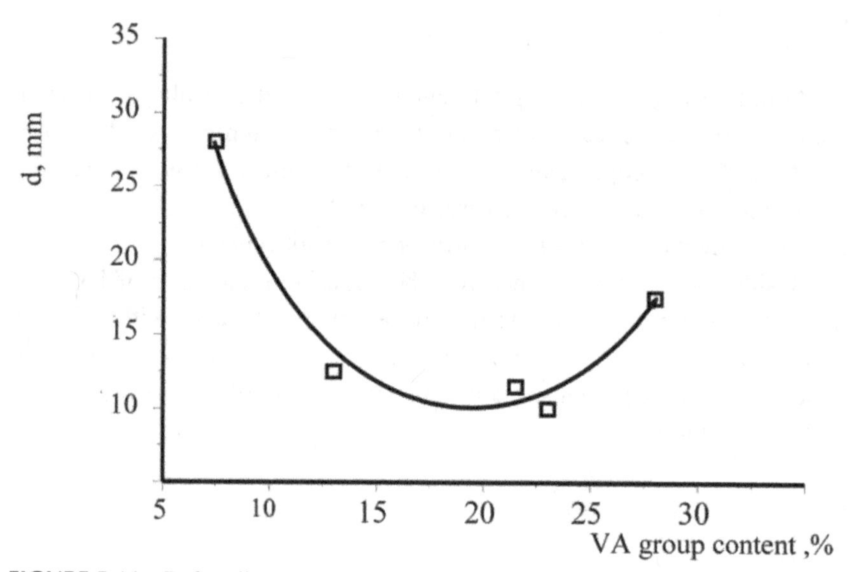

FIGURE 5.11 Defect diameter upon cathodic disbandment of EVA-based coatings as a function of the VA group content.

A non-linear pattern of the dependence of acid-base properties on the VA group content explicitly predicts both a non-linear nonmonotonic dependence of the relative acidity parameter (Fig. 5.10) and, according to the acid-base approach, the inverse dependence of defect diameter of EVA-based coatings upon cathodic disbandment (Fig. 5.11). By combining these two dependencies as a conventional d–ΔD plot, the dependence visualized in Fig. 5.12 is obtained to evidence again that the enhancement of acid-base interaction of adhesive and adherend facilitates the increase of adhesive joint strength.

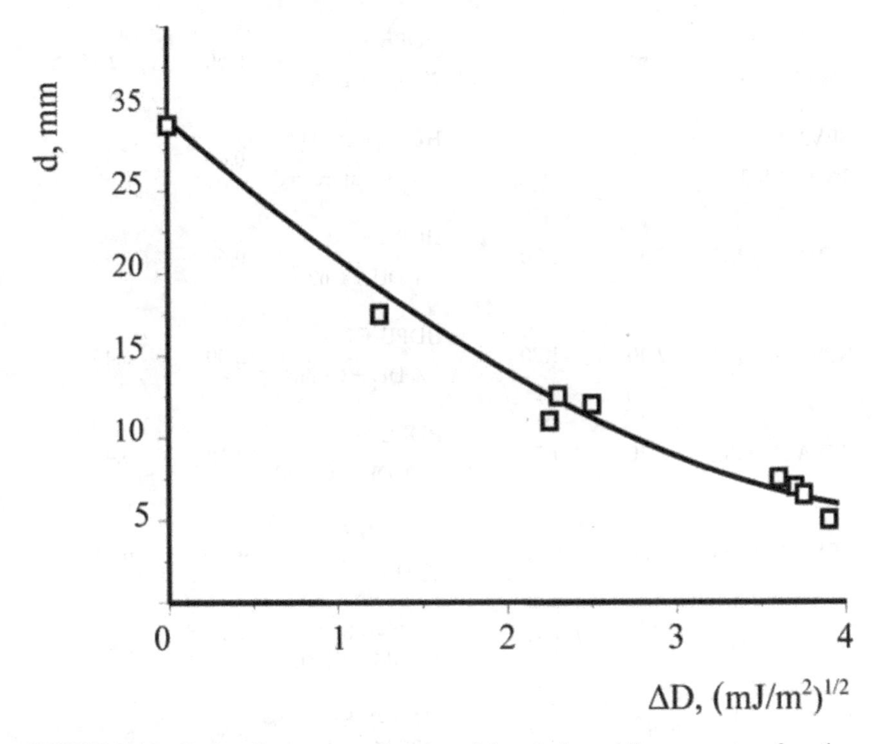

FIGURE 5.12 Defect diameter as a function of the relative acidity parameter of coatings based on EVA containing various VA group amount on St3 steel.

Experimental data for all metal – polyolefin coating systems studied are summarized in Table 5.2. Besides the modified polyethylenes and EVAs containing various VA group amount, PI-modified EVAs are also included.

TABLE 5.2 Acidity parameters of the components of polyolefin-based adhesive joint.

Adhesive joint	$D_{polymer,}$ $(mJ/m^2)^{1/2}$	$D_{metal'}$ $(mJ/m^2)^{1/2}$	Adhesive joint	$D_{polymer}$ $(mJ/m^2)^{1/2}$	$D_{metal'}$ $(mJ/m^2)^{1/2}$
SEVA 22 + 6% PI + St3	–4.00	1.70	HDPE + 2% DH+G65	0.60	4.55
SEVA 22 + 5% PI + St3	–3.75	1.70	HDPE + 2% DH + aluminum	0.60	–1.90
SEVA 22 + 4% PI + St3	–3.65	1.70	HDPE + 2% DH +D16	0.60	–2.15
SEVA 22 + 2% PI + St3	–3.60	1.70	HDPE + 2% DH +D16 heat treated	0.60	1.25
SEVA 22 + St3	–2.50	1.70	HDPE + 2% DH + ChZh1	0.60	6.45
SEVA 14 + St3	–2.40	1.70	HDPE + 1% DH + ChZh1	2.00	6.45
SEVA 20 + St3	–2.20	1.70	HDPE + 1.5%DH + ChZh1	2.10	6.45
SEVA 29 + St3	–1.20	1.70	LDPE-168 + ChZh1	4.05	6.45
SEVA 7 +St3	0	1.70	LDPE-168 + 2% MPDM + ChZh1	6.10	6.45
HDPE -289 + St3	2.30	1.70	LDPE-168 + 1.5% MPDM + ChZh1	6.50	6.45
HDPE slightly oxidized +St3	2.35	1.70	HDPE + 0.5% DH + ChZh 1	2.20	6.45

TABLE 5.2 *(Continued)*

Adhesive joint	$D_{polymer}$, $(mJ/m^2)^{1/2}$	D_{metal}, $(mJ/m^2)^{1/2}$	Adhesive joint	$D_{polymer}$ $(mJ/m^2)^{1/2}$	D_{metal}, $(mJ/m^2)^{1/2}$
HDPE + St3	3.60	1.70	LDPE-168 + 4% MPDM + ChZh 1	7.10	6.45
LDPE-168 +St3	4.05	1.70	LDPE-168+1% MPDM +ChZh1	8.00	6.45
HDPE + 2%DH + L90	0.60	8.40	HDPE + 2% DPP + titan	7.60	0.90
HDPE +1.5% DH + copper	2.10	3.30	LDPE-168 + 1% MPDM + titan	8.00	0.90

It is interesting that the dependence of defect diameter upon cathodic disbandment on ΔD for polyolefin coatings is similar to that for polyepoxide ones (Fig. 5.13).

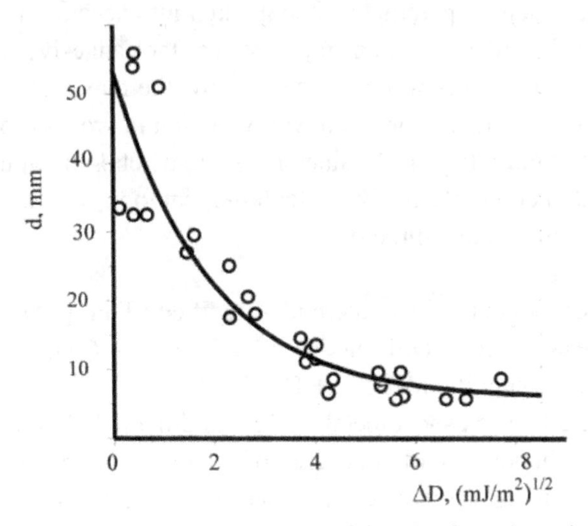

FIGURE 5.13 Defect diameter as a function of the ΔD of metal – polyolefin coating adhesive pairs:

In this case, a slight adhesive interaction is also observed in the range ΔD of 0–2.7 $(mJ/m^2)^{1/2}$ at a negligible difference in the acidity parameter values, since defect diameter in this range diminishes from the maximal observed down to ~15–20 mm (including the initial one), that is, it is e = 2.72 times less. Starting from the $\Delta D \approx 3$ and due to its further increase, defect diameter approaches to the initial one, which means no disbandment and, hence, a strong adhesive interaction. The generalized dependence is given by the exponent similar to that in the Eq. (5.2):

$$d=d_0+ A\exp(-\beta\Delta D)$$

where the parameters d_0, A and β have the same physical meaning as those for the dependence shown in Fig. 5.7 and are equal to 5 mm, 48 mm, and ~0.5 correspondingly.

Both variable time ranges of cathodic disbandment test and substantial differences in physical-mechanical properties of polyepoxide and polyolefin coatings (such as, residual stress level) do not allow generalizing the d–ΔD dependence for all systems studied. At the same time, similar pattern of this relationship for materials of different structure points to the versatility of the acid-base approach.

The possibility to predict adhesive interaction of adhesive with adherend, with their acidity parameters being taken into account, is verified by experiments. In order to obtain the most resistant adhesive interactions, substrates and adhesives were chosen to show the maximal difference in acidity parameters. In combination with variation in process conditions of sample formation, a high DD values for systems chosen stipulate a high adhesive interaction (defect diameter after cathodic disbandment happened to be equal to the initial one).

Samples of systems:

Titan (heat treated) – bis-maleimide-modified LDPE (DD = 6.65);

Titan (heat treated) – DPP-modified HDPE (DD = 6.3);

Brass – PAA-modified HDPE (DD = 7.8);

ChZh1 grade black sheet metal – PAA-modified HDPE (DD = 5.8).

Hydrogen bonds (a particular case of acid-base interactions), the energy of which is usually greater than that of van der Waals bonds, can be formed in adhesive joints across the interface. Their existence was truly verified and exemplified by many systems [213], including those discussed in present chapter. Thus, in case of modifying HDPE-based coatings by additives containing active functional groups (diphenylolpropan and PAAs) the adhesive interaction was found to increase substantially [102].

The effect observed is most probably due to hydrogen bonds which are formed between modifier and thermo-oxidized HDPE groups (on the one hand) and metal hydroxides (on the other hand). PAA forms hydrogen bonds with both oxygen-containing groups resulted from thermal oxidation of polymer (R_1 and R_2 are hydrocarbon residues) and active functional groups of acidic substrate. Possible mechanism of hydrogen bond formation is demonstrated in Figs. 5.14 and 5.15.

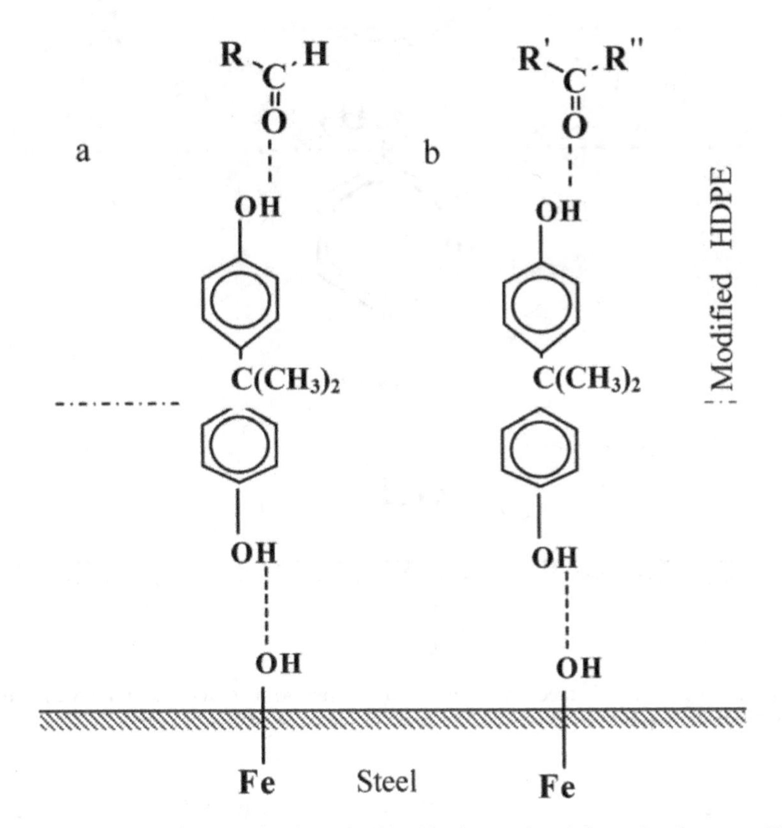

FIGURE 5.14 Possible mechanisms (a, b) of hydrogen bond formation between DPP-modified HDPE and St3 steel R, R,' R" are HDPE hydrocarbon residues.

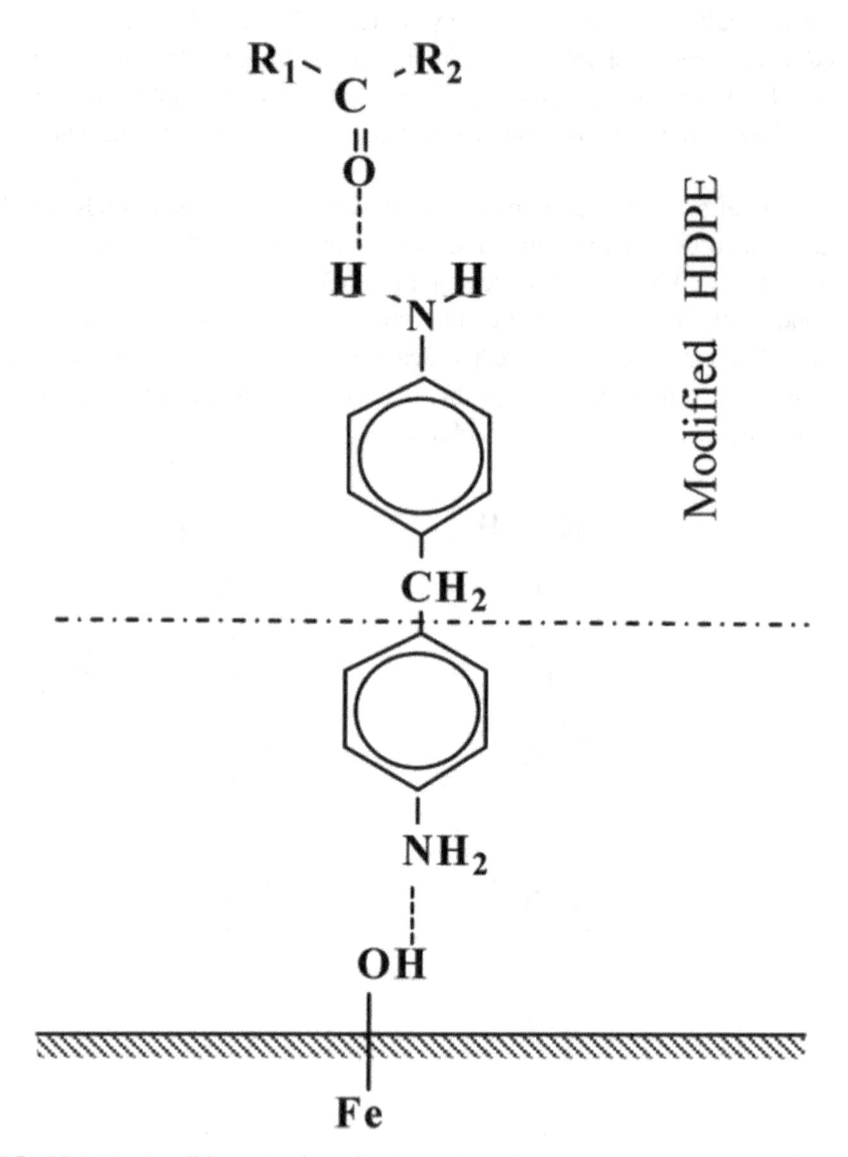

FIGURE 5.15 Possible mechanism of hydrogen bond formation between PAA-modified HDPE and St3 steel.

The results were verified by IR-spectroscopy used for model systems consisting of PAAs and their mixtures with fine iron, heat treated under process conditions for coating formation. IR-spectra of PAAs and their mixtures with iron profoundly differ (Fig. 5.16 a, b).

PAA-iron systems show changes in absorption bands at 3300–3500 cm^{-1} due to stretching vibrations of the N-H bond of primary amino groups. Changes include band broadening and absorption enhancement in low-frequency region, which indicates the formation of donor-acceptor bonds between the amino group and functional groups on iron surface.

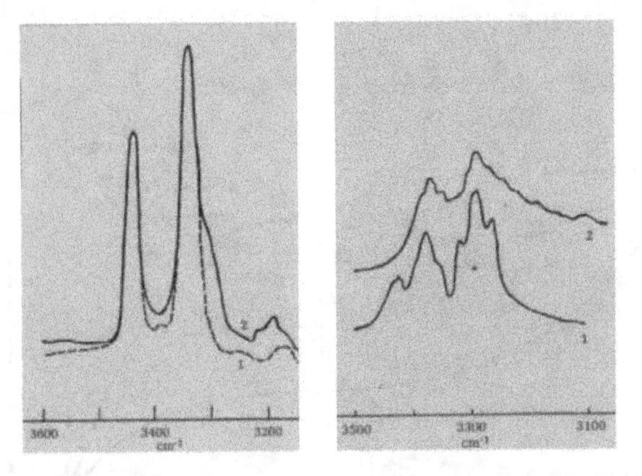

FIGURE 5.16 IR-spectra of PAAs (curve 1) and their mixtures with fine iron (curve 2): a – DH modifier; b – PAT polyamine.

5.4 THE RELATIONSHIP BETWEEN ACID-BASE AND ADHESIVE PROPERTIES OF POLYOLEFIN MIXTURES

Let us return to explanation of the fact of increase in adhesive characteristics of OREVAC grade ethylene-vinyl acetate-maleic anhydride terpolymers in the mixture with different EVAs (Section 4.4, Chapter 4). Thus, in case of coating based on SEMA 9307 + SEVA 20–20 system in the region of EVA concentrations of 70%, the experimental adhesive force dependence has the extreme and runs 4 kN/m higher than additive values are drawn (Fig. 5.17, Curve 1).

Similar dependence is observed in case of SEMA 9305 + SEVA 28–05 system (Fig. 5.18, Curve 1). For SEMA 9305 + SEVA 20–20 system this effect takes place at EVA concentration of 80%. In the last case, the experimental adhesive force value is greater than that of an individual terpolymer, which allows a synergistic effect of polymers in mixture on increasing the adhesion to steel to be assumed (Fig. 5.19, curve 1).

It is worthwhile noting that physical-mechanical characteristics of samples (tensile stress, tensile modulus, etc.) did not substantially deviate from additive values. Studies showed that percentage variation in the mixtures of polymers affects their surface acidity values (Table 4.11). The correctness of the results obtained became explainable due to using the acid-base approach.

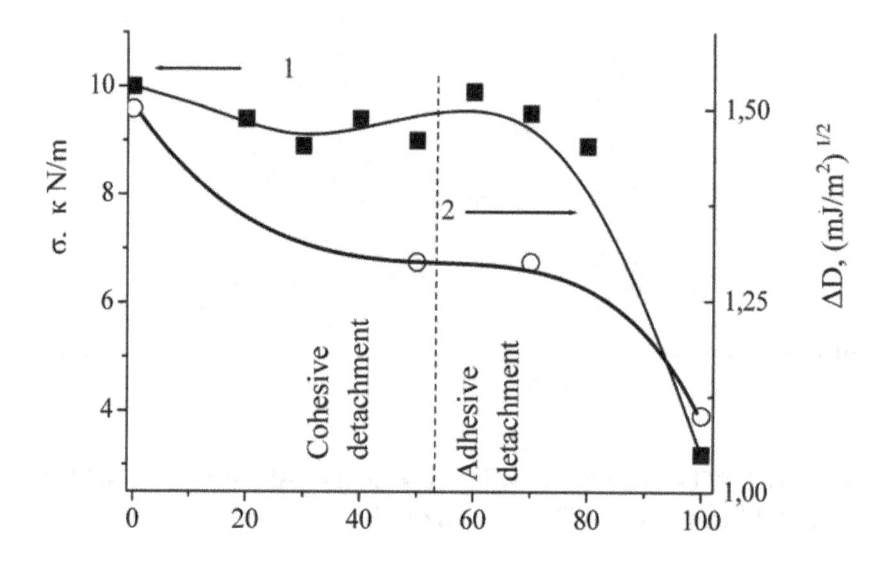

FIGURE 5.17 Adhesive force (Curve 1) and the relative acidity parameter (curve 2) as a function of SEVA 20 content in SEMA 9307-based composition.

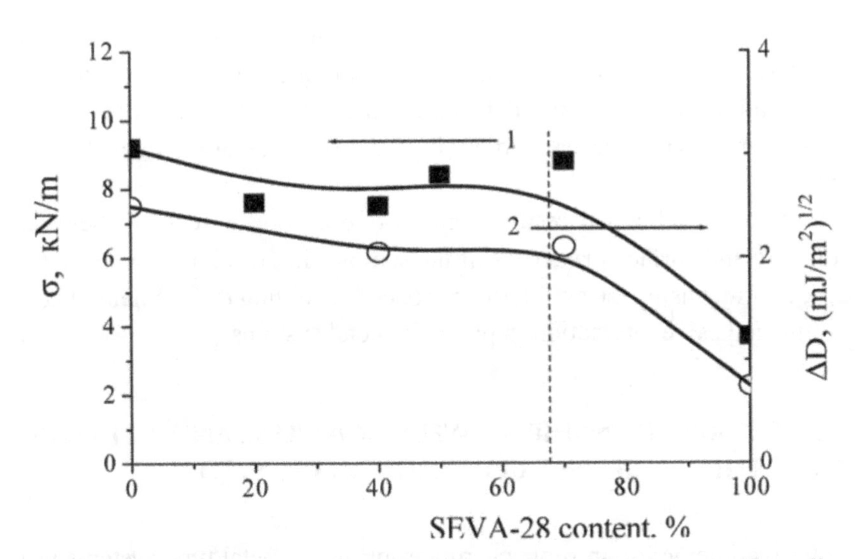

FIGURE 5.18 Adhesive force (Curve 1) and the relative acidity parameter (Curve 2) as a function of SEVA 28–05 content in SEMA 9305-based composition.

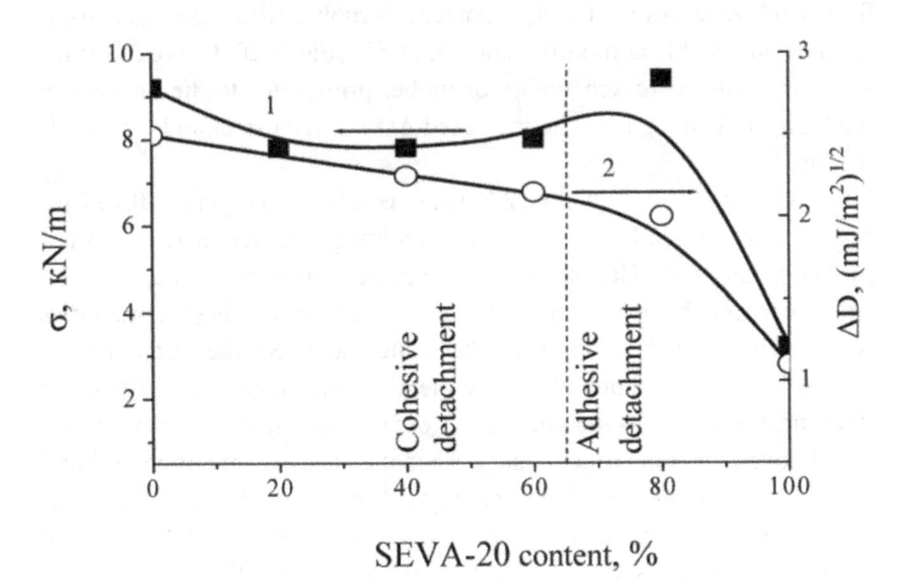

FIGURE 5.19 Adhesive force (Curve 1) and the relative acidity parameter (Curve 2) as a function of SEVA 20–20 content in SEMA 9305-based composition.

Within the area of adhesive detachment of compositions from steel, adhesive and acid-base characteristics are in agreement, namely, the adhesive force upon adhesive detachment increases as the relative acidity parameter rises in all cases studied (Figs. 5.17–5.19, Curves 1 and 2) [192, 199].

Again, the obtained results demonstrate a good agreement between adhesive and surface properties of the samples in question as well as evidence correct using the acid-base approach for solving the problem of controlling adhesive interaction in polymer-metal systems.

5.5 THE RELATIONSHIP BETWEEN ACID-BASE AND ADHESIVE PROPERTIES OF RUBBER COATING – METAL SYSTEM

Adhesive interaction in rubber – rubber primer – metal type systems was estimated by the peeling method at 180°. It was found that the force of peeling of a tape composed of both vulcanized adhesive and primer layers from steel increases as the PQD content in rubber rises. This increase is the most noticeable at modifier content of 5% (Fig. 5.20, Curve 2). At the same time, the decreased acidity of rubber primer due to the increase of the PQD content leads to the increased ΔD between steel and composite (curve 1).

In the absence of PQD, the ΔD value is rather small and is 0.15 $(mJ/m^2)^{1/2}$. As PQD is added, the relative acidity parameter increases, since surface acidity falls. Hence, the force of peeling rises from 20 up to 36 N/cm. It is reasonable to assume that such an increase of adhesive properties is caused by intensification of acid-base interaction. So, the greater primer surface basicity the more noticeable the acid-base interaction with acidic steel surface as well as the greater the force of peeling of rubber samples.

Of course, it is incorrect to suppose that adhesive force of chlorobutyl rubber composition – steel joint is only defined by acid-base interactions in the system. As repeatedly mentioned above, the force of adhesive joint comprises both adhesive and deformation constituents. The latter depends on mechanical properties of adhesive.

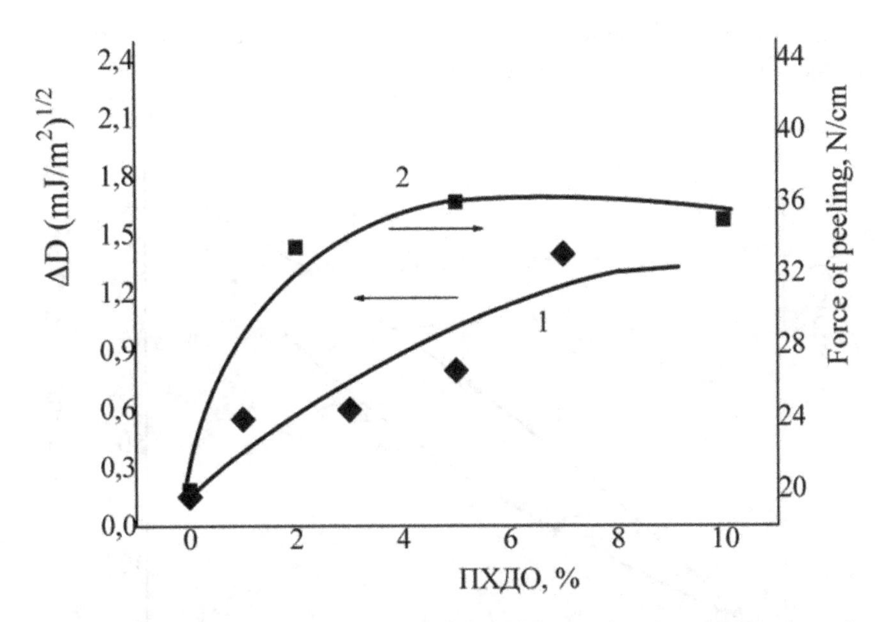

FIGURE 5.20 The ΔD value and the force of peeling as a function of PQD content in rubber composition.

The increase of adhesive properties of coating upon modifying rubber composite by manganese dioxide (the acidity parameter variation due to modification is shown in Fig. 4.8) is well explained by the acid-base theory (Fig. 5.21). Again, the increase in the force of peeling (curve 2) proceeds symbiotically as the relative acidity parameter rises (curve 1).

Finally, the theory in question allowed explanation of processes occurring upon modifying rubber by Acetonanyl grade TDQ in combination with PPR. As the Acetonanyl content in the primer layer increases, the force of peeling of the vulcanized tape from steel varies through the extreme, with the maximum being at 10% of Acetonanyl, and the breaking strength being 6.9 kg/cm. In this case, both dependences of the relative acidity parameter and the force of peeling on modifier content are in a well agreement again (Fig. 5.22). The greater the ΔD value of rubber-metal adhesive joint, the greater the adhesive force of this joint. As Acetonanyl is added without PPR, the interaction of adhesive with steel enhances slightly (Table 5.3) because of a low solubility of modifier in rubber. The acidity of surface is rather high at 5 and 10% of Acetonanyl, but it reduces as modifier content continues to rise.

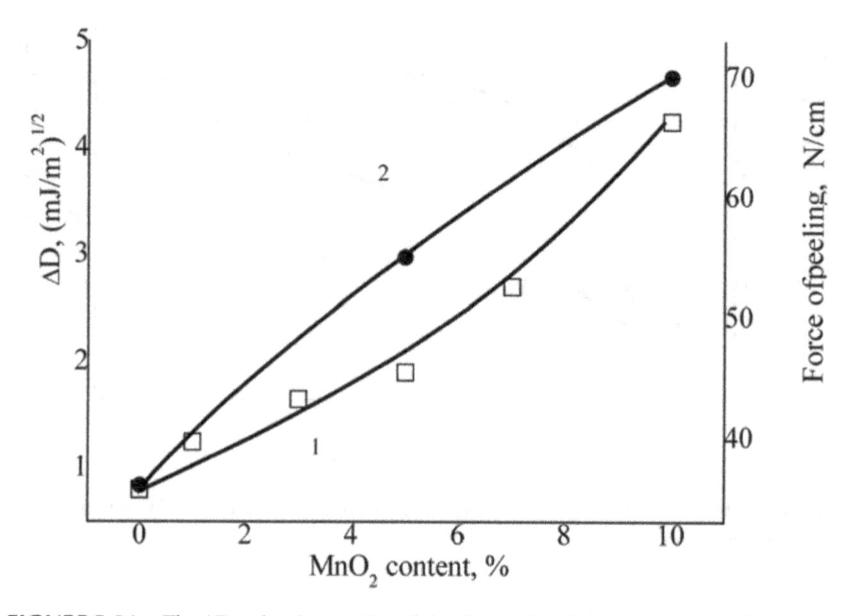

FIGURE 5.21 The ΔD value (curve 1) and the force of peeling (curve 2) as a function of manganese dioxide content in composition.

Thus, at 40% of Acetonanyl in composition without PPR the acidity parameter is 2 $(mJ/m^2)^{1/2}$ only. A film prepared from Acetonanyl solution shows D = –0.85 and, hence, is near basic.

In case of combining Acetonanyl with PPR, the major role in providing adhesive contact is probably played by acid-base interactions.

The reasons of a high value of the force of peeling upon introducing PPR together with Acetonanyl will be analyzed further in view of the acid-base approach.

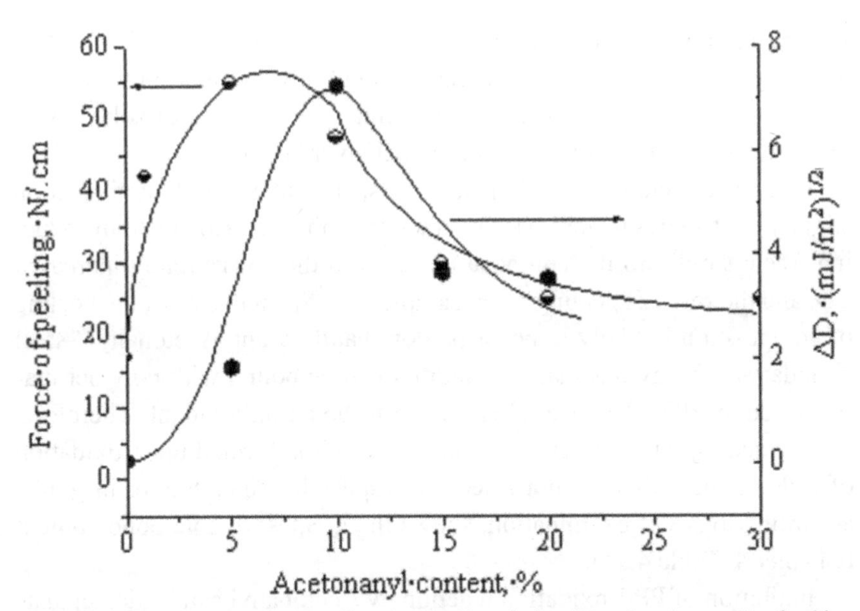

FIGURE 5.22 The force of peeling (curve 1) and the ΔD value (curve 2) of rubber composition as a function of Acetonanyl grade TDQ content.

TABLE 5.3 Acid-base properties and the force of peeling of rubber adhesive from steel.

№	Primer composition *, wt.—%			γ_s^d mJ/m²	γ_s^{ab} mJ/m²	D (mJ/M²)$^{1/2}$	DD	Force of peeling, N/cm at 20°C
	CIIR	PPR	Aceto-nanyl					
1	87	–	–	23.3	4.6	3.6	1.1	13
2		Acetonanyl		40.4	9.8	-0.84	3.34	–
3	82	–	5	24.7	12	7.5	5.0	17
4	77	–	10	26.4	11.1	7.7	5.2	21
5	67	20	–	22.3	9.7	1.3	1.2	33
6	57	30	–	31.2	2.9	1	1.5	35
7	52	30	5	21.6	3.0	6.5	4.0	55
8	47	30	10	32.4	2.4	8.5	6.0	47.5

*All compositions contain 10% of talc, 1% of zinc oxide, and 2% of ZDEC.

Introducing Acetonanyl into chlorobutyl rubber results in the substantially increased ΔD between rubber layer and steel, with the adhesion to steel being, however, increased slightly (Table 5.3, samples # 3, 4). Most probably, the reason is low Acetonanyl solubility in chlorobutyl rubber, namely 0.1%. Having occurred at the boundary of metal, undissolved Acetonanyl particles form unstable interfacial layer [169].

Due to its acidity according to Lewis, the rubber–PPR composition slightly contributes to acid-base interaction with steel. However, introducing Acetonanyl into the composition leads to the substantially increased ΔD, and the force of peeling rises (samples # 7, 8). In this case, the peeling of composition is likely to occur predominantly along Acetonanyl –steel boundary. It is obvious that the modifier can be bound with polymer matrix through PPR. Acetonanyl is known to be an antioxidant. Therefore, its interaction with the peroxide radical, which is formed upon oxidation of PPR olefin carbon, is not ruled out, especially since this resin grade, according to NMR examination, shows the greatest double bond content (Chapter 4, Table 4.11).

Inhibition of PPR oxidation reaction by Acetonanyl can be depicted as follows:

$$RH + O_2 \rightarrow R\bullet + HOO\bullet$$

$$R\bullet + O_2 \rightarrow ROO\bullet$$

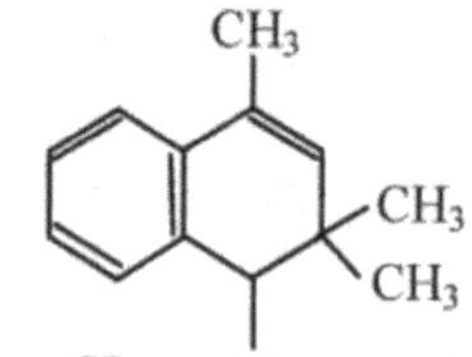

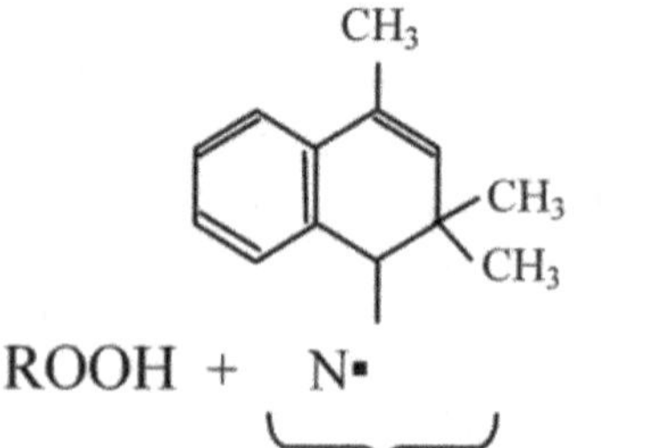

$$ROO\bullet + H-N \rightarrow ROOH + N\bullet \tag{5.4}$$

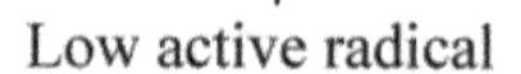

Low active radical

The aforesaid was verified by IR examinations of both PPR and Acetonanyl model solutions in carbon tetrachloride (Fig. 5.23). The spectrum of the product demonstrates broadening and shifting of the amino group peak at δ=3.35 ppm to low chemical shift region by 0.25 ppm (compared with Acetonanyl spectrum). Probably, PPR oxidation reaction is inhibited at the stage of hydroperoxide formation. However, small amounts of antioxidants do really act, so Acetonanyl, when then added, is not consumed, but most probably travels onto surface making it less acidic, since the modifier itself is basic due to amino groups. Thus, Acetonanyl is able to undergo acid-base interaction with steel and inhibit PPR oxidation. In its turn, the resin, when dissolved in polymer matrix, provides indirect interaction of Acetonanyl with rubber.

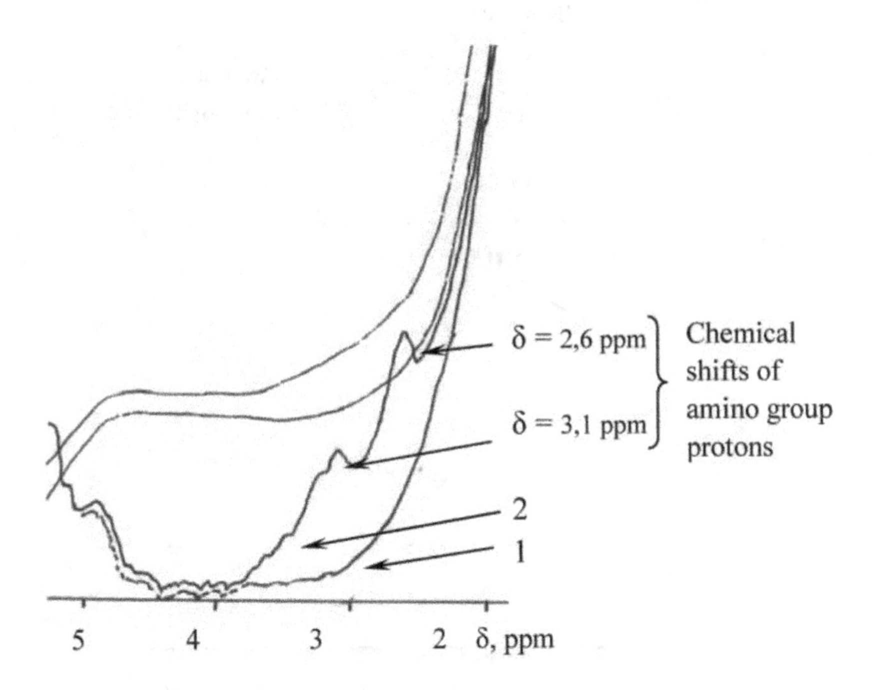

FIGURE 5.23 NMR spectra of PPR (1) and PPR–Acetonanyl reaction product (2).

The studies carried out allowed the model of acid-base interaction for the system considered to be proposed (Fig. 5.24). The figure reflects the fact that inhibition of PPR oxidation by Acetonanyl at the composition material – air interface seems to result in the formation of carboxyl groups which make surface pronouncedly acidic, more acidic than in the absence of antioxidant. At metal boundary, due to the lack of oxygen supply, Acetonanyl is bound with both PPR and oxidized steel by donor-acceptor interaction. Again, it is necessary to note that the case considered, rather probable as it may seem, is only a variant of interactions possible.

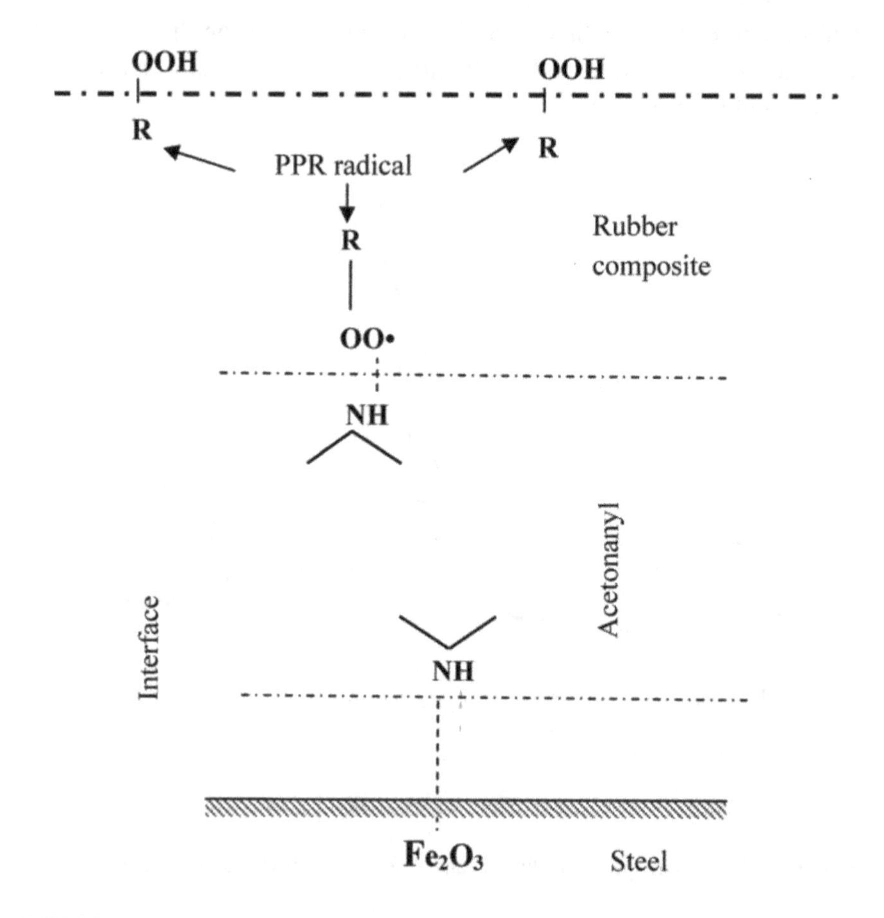

FIGURE 5.24 Possible model of acid-base interaction of Acetonanyl -and-PPR-modified rubber with St3 steel.

So, variation of the force of peeling of compositions from steel is observed upon modification in all systems studied together with variation of surface acidity. All experimental data including compositions and acidity parameters of chlorobutyl rubber based adhesive joints in question are summarized in Table 5.4. Figure 5.25 depicts the dependence of the force of peeling of rubber compositions using different vulcanizing systems on the relative acidity parameter ΔD between adhesive and metal.

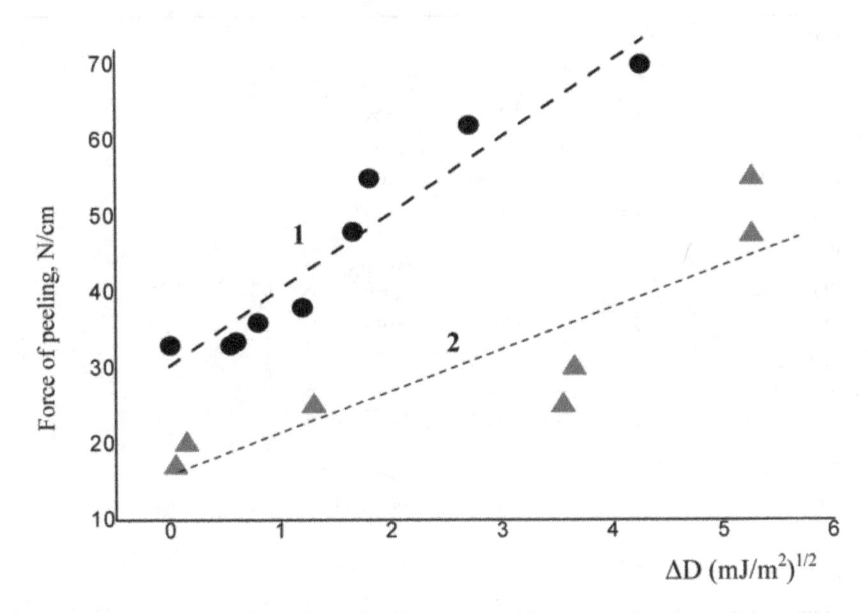

FIGURE 5.25 Force of peeling of rubber compositions as a function of the relative acidity parameter (generalized dependences).

Two data groups are presented in the figure. PQD was used as the vulcanizing agent for samples of the first group (curve 1), whereas zinc oxide and talc were used as the vulcanizing system for samples of the second group (curve 2). According to their physical-mechanical properties, composition materials of our interest are low-modulus systems. The role of the deformation constituent is known to be rather noticeable upon estimating the adhesive force of such composites. PQD is more efficient vulcanizing agent compared with zinc oxide – talc system, which leads to the difference in physical-mechanical of compositions and, hence, to various contributions of the deformation constituent to adhesive force. Thus, we observed various forces of peeling for the composites using different vulcanizing systems at the same ΔD values.

TABLE 5.4 Composition and acidity parameters of rubber based adhesive joints.

Composition	$D_{polymer}$, $(mJ/m^2)^{1/2}$	D_{metal}, $(mJ/m^2)^{1/2}$	Composition	$D_{polymer}$, $(mJ/m^2)^{1/2}$	D_{metal}, $(mJ/m^2)^{1/2}$
Group 1					
PPR-20%, PI-5%, PQD-5%	3.3	3.3	PPR -25%, PI -5%, ZnO-2%, talc -10%	3.15	3.3
PPR -25%, PI -5%, PQD -1%	2.75	3.3			
PPR -25%, PI -5%, PQD -2%	2.7	3.3	PPR -30% ZnO-2%, talc -10%	1.3	1.25
PPR -25%, PI -5%, PQD -5%	2.5	3.3			
PPR -25%, PI -5%, PQD -5%, MnO_2-1%	2.1	3.3	Acetonanyl –40% ZnO-2%, talc -10%	2.0	3.3
PPR -35%, PI -5%, PQD -5%	4.75	3.3			
PPR -25%, PI -5%, PQD -5%, MnO_2-3%	1.65	3.3	PPR -30%, Acetonanyl -20% ZnO-2%, talc -10%	4.8	1.25

TABLE 5.4 *(Continued)*

Composition	$D_{polymer}$, $(mJ/m^2)^{1/2}$	D_{metal}, $(mJ/m^2)^{1/2}$	Composition	$D_{polymer}$, $(mJ/m^2)^{1/2}$	D_{metal}, $(mJ/m^2)^{1/2}$
PPR -25%, PI -5%, PQD -5%, **MnO_2 5%**	1.5	3.3	PPR -30%, Acetonanyl -15% ZnO-2%, talc -10%	4.9	3.3
PPR -25%, PI -5%, PQD -5%, **MnO_2 7%**	0.6	3.3	PPR -30%, Acetonanyl -10% ZnO-2%, talc -10%	8.5	1.25
PPR -25%, PI -5%, PQD -5%, MnO_2 10%	–0.95	3.3	PPR -30%, Acetonanyl -5% ZnO-2%, talc-10%	6.5	1.25

The results obtained confirm the applicability of the acid-base approach to rubber adhesive – metal systems. Determination of surface characteristics allows studying the effect of any modifier on surface properties of composition and estimating optimal concentrations in formulations for further prediction and targeted control of interaction at metal – primer and primer – rubber adhesive interfaces.

5.6 THE RELATIONSHIP BETWEEN ACID-BASE AND ADHESIVE PROPERTIES OF MODIFIED RUBBER MIXES

A confirmation of the fact that acid-base interactions play a role in providing a strong contact between rubber adhesive and metal is exemplified by acid-base and adhesive properties of brass and isoprene rubber mixes modified by different additives which alter the acidity parameter of vulcanized rubber surface in broad ranges. In particular, these systems simulate interfacial interaction of breaker vulcanizate and brass-plated steel cord in tyres. Presently, the problem of adhesion in similar systems is being considered mostly from formulation-processing point of view, with acid-base consideration being ignored. Examination of the role of acid-base interactions played in vulcanizate – brass-plated steel cord systems as well as their effect on the strength of vulcanizate – steel cord contact is, therefore, of a great interest.

Multiple studies and practical application showed a high efficiency of vulcanizate – steel cord bonding through brass coating and simultaneous using adhesion promoters in elastomer compositions [214–216]. The efficiency of such modifiers as organic cobalt salts is connected with their positive effect on the formation of a thin film of non-stoichiometric copper sulfide on brass surface used as adhesive as well as the formation of strong chemical bonds and labile physical bonds at the interface. Composition and surface-energy characteristics of samples are presented in Tables 4.15 and 4.16, Chapter 4. Acid-base properties of brass surface will change, if the coating is peeled off (Table 5.5). A sharp increase in basicity is caused by brass sulfidation during vulcanization [214].

TABLE 5.5 Brass surface characteristics.

Surface	γ_s^d mJ/m^2	γ_s^{ab} mJ/m^2	γ_s mJ/m^2	D (mJ/m^2)$^{1/2}$
Brass	23.1	13.3	36.4	4.0
Sulfidated brass	25.4	13.2	38.6	–3.0

A key role acid-base interactions play in the formation of bonds between modified vulcanizate and brass was verified by the correlation between the ΔD value of adhesive joints and the resistance to cathodic disbandment, which is taken as a conditional measure of adhesion symbolized as boxes in Fig. 5.26. The same figure contains data on cathodic disbandment of rubber adhesives from steel symbolized as circles. As this method allows minimizing the deformation constituent of adhesive force, the plot is general in spite of various physical-mechanical characteristics of samples. Empirically chosen dependence (a solid curve, Fig. 5.26) is well described by the equation mathematically similar to the equation 5.2.

Thus, the plotted graph is described by the following equation:

$$d = 5+66\exp(-\Delta D/3) \tag{5.5}$$

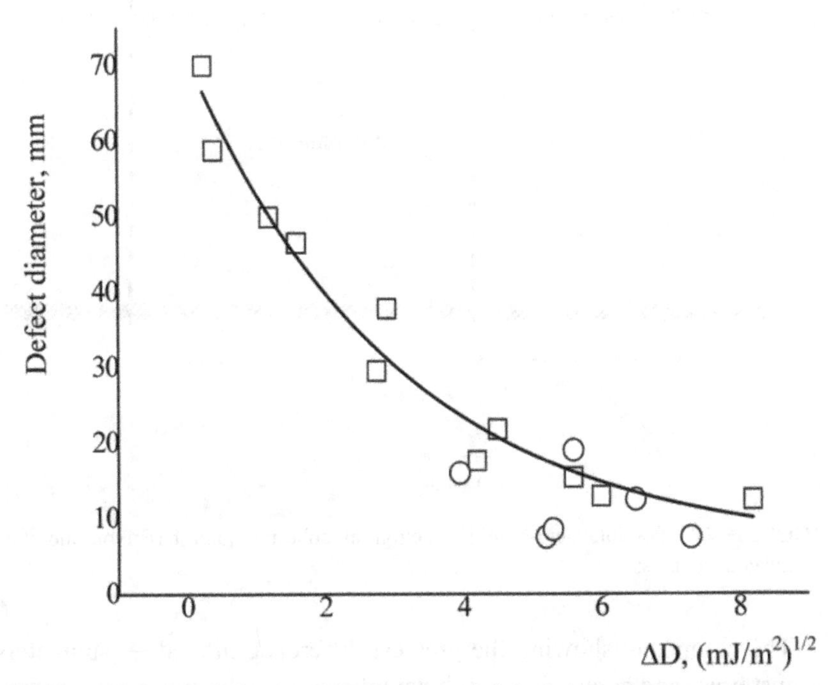

FIGURE 5.26 Defect diameter as a function of the relative acidity parameter upon cathodic disbandment for different adhesive systems.

The increase of vulcanizate adhesive properties will be observed, if vulcanizate surface acidity enhances. As the brass sulfidated during rubber mix vulcanization is basic ($D = -3.0$ $(mJ/m^2)^{1/2}$), the increase in the resistance to cathodic disbandment due to the growth of the relative acidity parameter is the evidence of the role of acid-base interactions played in the formation of adhesive bonds in vulcanizate – steel cord system [173–176]. The mechanism of such an interaction implies the generation of acid-base bonds between acidic active functional groups of vulcanizate surface (such as, metal atoms of adhesion promoters) and basic acceptor centers, such as oxygen or sulfur atoms of metal substrate. Figure 5.27 depicts possible interactions in modified vulcanizate-based adhesive joint.

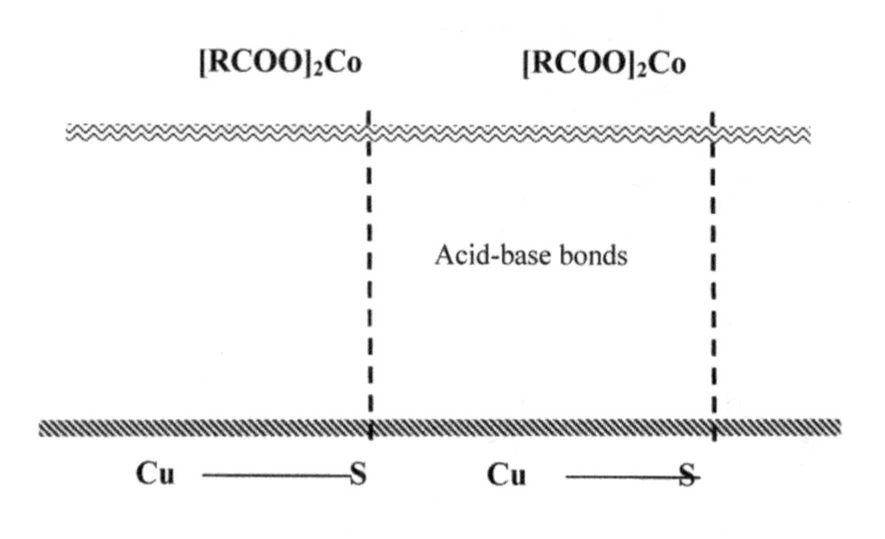

FIGURE 5.27 Possible acid-base interaction in adhesive joint based on modified vulcanizate and brass.

The composite showing the greatest difference in acidity parameters against brass and containing cobalt naphthenate as adhesion promoter was taken here as a basis. According to the reported data, brass surface may also contain copper and zinc oxides, but in this case, Figure 5.27 depicts copper sulfides because it is copper compounds that play a key role in the formation of vulcanizate – brass adhesive contact [214–216]. The result obtained demonstrates that using the acid-base approach to solving the problem of adhesive interaction control in breaker vulcanizate – steel cord systems offers the promise.

The studies performed allowed formulating epoxy, polyolefin, and rubber compositions showing the best adhesive ability to different metals. Adhesive compositions optimized by using the acid-base approach and successfully passed laboratory tests at different enterprises of Russia and the Republic of Tatarstan were formulated.

KEYWORDS

- **adhesive and adherend properties**
- **adhesive technologies**
- **ethylene-acrylic acid copolymer**
- **metal-polymer systems**
- **polymer-polymer systems**
- **relative acidity parameter**
- **vulcanizate**

CONCLUSION

Finally, it is worthwhile noting that the acidity parameter is an important and informative characteristic. The studies carried out showed that this parameter adequately reflects acid-base properties of surface and is very sensitive to any variation in processing and formulation of polymer compositions. The D values obtained are natural, explainable, and verified for the systems the surface acidity of which may be predictable basing on the known chemical composition. The knowledge of the acidity parameter value can be of special importance in case of surfaces of complicated composites containing five-to-ten and more modifiers. To predict the nature of such surfaces *a priori* seems difficult, and therefore, measuring the D value will allow information for using them in adhesive joints of different purpose.

Today, it is important that the acidity parameter be determined correctly for any solid smooth surface, affected in order to enhance adhesive interaction in polymer – metal systems, and applied for predicting interfacial interaction in real systems of interest. These problems can be successfully solved by using the acid-base approach. This was verified by studies performed, targeted to solving the problem of the enhancement of adhesive interaction between phases in contact. The results obtained may be useful in designing specific coating-on-metal systems using rubbers, polyolefins, and polyepoxides.

BIBLIOGRAPHY

1. Fowkes, F. M. (1990). J. Adhes. Sci. Technol. 4(8), 669–691.
2. Della Volpe, C. et al. (2003). Recent theoretical and experimental advancements in the application of van Oss – Chaudury – Good acid–base theory to the analysis of polymer surfaces, I. General aspects; J. Adhesion Sci. Technol. 17(11), 1477–1505.
3. Kinlock, E. (1991). Adhesion and adhesives. Science and technology. - M.: Mir, 484 p.
4. Acids and bases. (1963). A short chemical encyclopedia, - M.: Sovetskaya entsiklopediya. 2, 581–587.
5. Brensted J. N. (1923). Rec. Trav. Chim. Pay-Bas. 42, p.718
6. Lewis, G. N. (1923).Valence and the Structure of Atoms and Molecules. - New York: Chemical Cataloguing Co., p.142.
7. Shatenstein, A. I. (1949). Acid–base theories, M.L.
8. Izmailov, N. A. (1967). Selected works; Yatsimirsky K. B. (Ed.). Kiev: Naukova Dumka, 460 p.
9. Shvarts, M. (1970). Ions and ion pairs; Uspekhi khimii. 39(7), 1260–1275.
10. Diogenov, G. G. (1984). The criterion of acid–base properties of oxides; Izvestiya vysshykh uchebnykh zavedeny. Khimiya i khimicheskaya tekhnologiya. 27(10), 1131–1134.
11. Pearson, R. G. (1973). Hard and Soft Acids and Bases. Dowden, Hutchinson and Ross, Stroudsburg, PA.
12. Pearson, R. G., & Songstadt, J. (1967). J. Org. Chem. 32, 2899.
13. Usanovich, M. I. (1970). Studies on the theory of solutions and the acid–base theory: Selected works; Sumarokova, T. N. (Ed.). Alma-aty: Nauka, 364 p.
14. Pankratov, A. N. (2006). Acids and bases in chemistry / Saratov University Printing Office, 196 p.
15. Milliken, R. S. (1952). J. Phys. Chem. 56, p.801.
16. Johnson, K. L., Kendall, K., & Roberts, A. D. (1971). Proc. Roy. Soc. vol. A324, p.301.
17. Fowkes, F. M., & Mostafa, M. (1978). Jnd. Eng. Chem. Prod. Res. Dev., v.17, p.3.
18. Rance, D. G. (1985). In: Industrial Adhesion Problems. Ed. D. M., Brewis, D. Briggs. – Oxford: Orbital Press. p.48.
19. Fowkes, F. M. (1964). Ind. Eng. Chem. 56(12), 40–52.
20. Oss, C. J., van, Good, R. J., & Chaudhury, M. K. (1988). Langmuir. 4, 884–891.
21. Fowkes, F. M. (1987). Role of acid–base interfacial bonding in adhesion; J. Adhesion Sci. Tech. 1(1), 7–27.
22. Deryagin, B. V., Krotova, N. A., & Smilga, V. P. (1973). Adhesion of solids.; M.: Nauka, 279 p.
23. Zimon, A. D. (2001). Colloid Chemistry: Teaching aid; Zimon, A. D., Leshchenko N. F., – M.: AGAR, 320 p.
24. Young, T. (1805). Trans. Roy. Soc. 95, p.65.
25. Eick, J. D., Good, R. J., & Neumann, A. W. (1975). J. Colloid Interface Sci. 53, p.235.
26. Oliver, J. F., Huh, C., & Mason, S. G. (1980). Colloids Surf. 1, 79.
27. Dettre, R. H., & Johnson, Jr. R. E. (1965). J Phys Chem. vol. 69, 1507.
28. Neumann, A. W., & Good, R. J. (1972). J. Colloid Interface Sci. 38, 341.
29. Wenzel, R. N. (1936). Ind Eng Chem. vol.28, p.988.

30. Summ, B. D. (1976). Physical-chemical basics of wetting and spreading; Summ, B. D., Goryunov, Yu.V. – M.: Khimiya, 232 p.
31. De Gennes, P. J. (1987). Uspekhi fis. nauk. 51(4), 619 p.
32. Gnusin, N. P., & Kovarsky, N.Ya. (1970). The roughness of electrodeposited surfaces: Teaching aid; Nauka, Novosibirsk, 236 p.
33. Starov, V. M. (1992). Adv Colloid Interface Sci. 39, p.147.
34. Long, J., Hyder, M. N., Huang, R. Y. M., & Chen, P. (2005). Adv Colloid Interface Sci. 118. p.173.
35. Zisman, W. A. (1964). Advances in Chemistry Series, 43 Ed. R. F. Gould Washington American Chemical Society. 1.
36. Tavana, H. (2007). Contact angle hysteresis on fluoropolymer surfaces; H. Tavana, D. Jehnichen, K. Grundke, M. L., Hair, A. W., Neumann; Advances in Colloid and Interface Science. No 134–135. 236–248.
37. Fadeev, A. Y., & McCarthy, T. J. (1999). Langmuir. 15, 3759.
38. Yasuda, T., Miyama, M., Yasuda, H. (1994). Langmuir. 10, 583.
39. Sedev, R. V., Petrov, J. G., & Neumann, A. W. (1996). J Colloid Interface Sci. 180, 36.
40. Hennig, A., Eichhorn, K. J., Staudinger, U., Sahre, K., Rogalli, M., & Stamm, M. (2004). Langmuir. 20, 6685.
41. Lam CNC, Wu, R., Li, D., Hair, M. L., & Neumann, A. W. (2002). Adv Colloid Interface Sci. 96, 169.
42. Kamusewitz, H., Possart, W., & Paul, D. (1999). Colloids Surf, A. 156, 271.
43. Della Volpe, C., Maniglio, D., Morra, M., & Siboni, S. (2002). Colloids Surf, A. 206, 47.
44. Andrieu, C., Sykes, C., & Brochard, F. (1994). Langmuir. 10, 2077.
45. Decker, E. L., & Garoff, S. (1996). Langmuir. 12, 2100.
46. Decker, E. L., Frank, B., Suo, Y., & Garoff, S. (1999). Colloids Surf, A. 156, 177.
47. Cassie, A. B. D. (1948). Discuss Faraday Soc. 3, 1120.
48. Fowkes, F. M., Riddle, F. L., Pastore, Jr., W. E., & Weber, A. A. (1990). Colloids Surf. No 43. 367–387.
49. Fowkes, F. M. (1967). Treatise on Adhesion and Adhesives / Ed. Patrick, R. L.; New York: Marcel Dekker, 1, 352.
50. Chibowski, E. (2007). On some relations between advancing, receding and Young's contact angles; Chibowski, E.; Advances in Colloid and Interface Science. No 133, 51–59
51. Fowkes, F. M. (1963). J. Phys. Chem. 67, 2538.
52. Fowkes, F. M. (1972). J. Adhesion. 4, 155.
53. Van Oss, C. J. (1994). Interfacial Forces in Aqueous Media; Dekker, New York. 1212.
54. Van Oss, C. J., Chaudhury, M. K., & Good, R. J. (1987). Adv. Colloid Interface Sci. 28, 35.
55. Good, R. J., & Girifalco, L. A. (1960). J. Phys. Chem. 64. 561.
56. Hildebrand, J. H., & Scott, R. L. (1950). Solubility of Nonelectrolytes, 3rd ed. Reinhold, New York.
57. Good, R. J., & Elbing E. (1970). Ind. Eng. Chem. 62(3), p. 54.
58. Good, R. J., Van Oss, C. J., & Chaudhury, M. K. (1988). Chem. Rev. v.88, p. 927.
59. London, F. H. (1937). Trans. Faraday Soc. v.33, p.8.
60. Berthelot, D. (1898). Compt. Rend. V.126, p.1703.
61. Keesom, W. H. (1921). Phys. Z. V.22, p.126.
62. Debye, P. (1921). Phys. Z. V.21, p.178.
63. Owens, D. K., & Wendt, R. C. (1969). J. Appl. Polymer Sci. 13, 1740.
64. Kaelble, D. H., & Uy, K. C. (1970). J. Adhesion. 2, 50.
65. Wake, W. C. (1978). Polymer, 19, P.291
66. Panzer, J. (1973). J. Colloid Interf. Sci. 44(2), 142.

67. Wu, S. (1973). In: Recent Advances in Adhesion, Ed. L. H., Lee. – London: Gordon and Breach, P.45
68. Van Oss, C. J., Chaudhury, M. K., & Good, R. J. (1989). Separation Sci. Technol. 24, 13.
69. Van Oss, C. J., Good, R. J., & Busscher, H. J. (1990). Dispersion Sci. Technol. 11, 77–81.
70. Good, R. J., Chaudhury, M. K., & van Oss, C. J. (1991). Fundamentals of Adhesion, L. H. Lee (Ed.). Ch. 3. New York, Plenum Press.
71. Good, R. J. (1992). J. Adhes. Sci and Technol. 6, 1269–1302.
72. Kollman, P., McKelvey, J., Johansson, A., & Rothenberg, S. (1975). J. Am. Chem. Soc. v.97, p.955.
73. Kollman, P. (1977). J. Am. Chem. Soc. V.99, p.4875.
74. Hobza, P., &Zaharadnik, R. (1980). Weak Intermolecular Interaction. Elsevier, New York, 171–176.
75. Milliken, R. S. (1952). J. Phys. Chem. 56, 801.
76. Drago, R. S., Vogel, C. G., & Needham, T. E. (1971). A Four-Parameter Equation for Predicting Enthalpies of Adduct Formation; J. Amer. Chem. Soc. 93(23), pp. 6014–6026.
77. Gutmann, V. (1978). The Donor-Acceptor Approach to Molecular Interactions. Plenum Press, New York, 134–136.
78. Swain, H. (1984). Private communication.
79. Joslin, S. T., & Fowkes, F. M. (1985). IEC Prod. Res. Dev. 24, p.369.
80. Fowkes, F. M., McCarthy, D. C., & Tischler, D. O. (1985). In: Molecular Characterization of Composite Interfaces, Plenum Press, New York. 401–411.
81. Fowkes, F. M., Dwight, D. W., Cole, D. A., & Huang, T. C. (1990). J. Non-Cryst. Solids 120, pp. 47–60.
82. Fowkes, F. M., Jones, K. L., Li, G., & Lloyd, T. B. (1989). Energy Fuels, 3, pp. 97–105.
83. Fowkes, F. M., Tischler, D. O., Wolfe, J. A., Lannigan, L. A., Ademu-John, C. M., & Halliwtll, M. J. (1984). J. Polym. Sci., Polym. Chem. Ed. 22, 547–566.
84. Kwei, T. K., Pearce, E. M., Ren, F., & Chen, J. P. (1986). J. Polym. Sci., Polym. Phys Ed. v.24, p.1597.
85. Valia, D. (1988). Ph.D. Thesis, Lehigh University.
86. Riddle, F. L., & Fowkes, F. M. (1990). J. Am. Chem. Soc. v.112, 3259–3264.
87. Casper, L. A. (1985). Ph.D. Thesis, Lehigh University.
88. Backstrom, K., Lindman, B., & Engstrom, S. (1988). Langmuir. 4, p.372.
89. Mason, J. G., Siriwardane, R., & Wightman, J. P. (1981). J. Adhesion 11, 315–328.
90. Lloyd, D. R., Ward, T. C., & Schreiber, H. P. (Eds), (1989). ACS Symposium Series Am. Chem. Soc., Washington, DC. 391, pp. 217–229.
91. Fowkes, F. M. (1985). in: Surface and Interfacial Aspects of Biomedical Polymers, J. D., Andrade (Ed.), Plenum Press, New York. V.12, pp. 337–372.
92. Schultz, J., & Lavielle, (1989). in: Inverse Gas Chromatography, D. R., Lloyd, T. C., Ward and, H. P., Schreiber (Eds), ACS Symposium Series; Am. Chem. Soc., Washington, DC. 391, p. 185.
93. Schreiber, H. P., Richard, C., & Wertheimer, M. R. (1983). in: Physicochemical Aspects of Polymer Surfaces, K. L., Mittal (Ed.), Plenum Press, New York. 2, pp. 739–748.
94. Anderson, H. R., Fowkes, F. M., & Hielcher, F. H. (1976). J. Polym. Sci., Phys. Ed. 14, 809.
95. Fowkes, F. M., & Harkins, W. D. (1940). J. Amer. Chem. Soc. v.62. 3377.
96. Vrbanac, M. D., & Berg, J. C. (1990). J. Adhes. Sci. Technol. 4(4), 255–266.
97. Della Volpe, C., & Siboni, S. (2000). Acid–base surface free energies of solids and the definition of scales in the Good – van Oss – Chaudhury theory; J. Adhesion Sci. Technol. 14(2), 235–272.

98. Fowkes, F. M., Kaczinski, M. B., Dwight, D. W., & Kelly, P. M. (1991). Langmuir. No 7, 2464.
99. Starostina, I. A. (1996). The role of primary aromatic amines in enhancing adhesive interaction of the modified polyethylene with steel; Author's abstract, Ph.D. thesis in Engineering, Kazan. – Kazan State Technol. Univ.
100. Starostina, I. A., The influence of the composition of epoxy formulations on energy, acid–base, and adhesive characteristics of coating surfaces; Starostina, I. A., Kustovsky, B. Ya., Garipov, R. M., Khuzakhanov, R. M., & Stoyanov, O. V. (2006). Lakokrasochnye materialy i ikh primenenie. No 8, 34–39.
101. Berger, E. J. (1990). J. Adhes. Sci. and Technol. 4(5), 373–391.
102. Starostina, I. A. (1998). The Role of Primary Aromatic Amines in Adhesion in Polyethylene-Steel Systems; Starostina, I. A., Stoyanov, O. V., Bogdanova, S. A., Deberdeev, R. Ya., Kurnosov, V. V., Zaikov, G. E.; Polymers & Polymer Composites. 6(8), 523–533.
103. Stoyanov, O. V. (1998). Changes in chemical structure of polyethylene coatings formed in the presence of a primary aromatic amine; Stoyanov, O. V., Starostina, I. A., Kurnosov, V. V., Deberdeev, R. Ya.; Zhurnal prikladnoi khimii. 11, 1871–1874.
104. Stoyanov, O. V. (1998). Interaction of primary amines with polyethylene; Stoyanov, O. V., Starostina, I. A., Kurnosov, V. V., Deberdeev, R. Ya.; Kuznetzov, A. M., Remizov, A. B.; Russian Polymer News. 3(2), 9–12.
105. Starostina, I. A. (1999). IR-Study of Polyethylene and Primary Aromatic Amines Interaction; Starostina, I. A., Deberdeev, R. Ya., Remizov, A. B., Kurnosov, V. V., Zaikov, G. E., Stoyanov, O. V.; Oxidation Commu-nications. 22(2), 171–177.
106. Starostina, I. A. (1999). The Role of Primary Aromatic Amines in the Intensification of Adhesion Interaction in Polyethylene-Steel System; Starostina, I. A., Stoyanov, O. V., Bogdanova, S. A., Deberdeev, R. Ya., Kurnosov, V. V., Zaikov, G. E.; Intern. J. Polymeric Mater. Vol.44, 35–51.
107. Starostina, I. A. (2001). Studies on the Surface Properties and the Adhesion to Metal of Polyethylene Coatings Modified with Primary Aromatic Amines; Starostina, I. A., Stoyanov, O. V., Bogdanova, S. A., Deberdeev, R. Ya., Kurnosov, V. V., Zaikov, G. E.; J. Appl. Pol. Sci. 79, 388–397.
108. Starostina, I. A. (2001). Modification of polyethylene by polyfunctional substances; Starostina, I. A., Stoyanov, O. V., Khuzakhanov, R. M., Kurnosov, V. V., Deberdeev, R. Ya.; Vestnik Kazanskogo Technol. Univ., 259–271.
109. Starostina, I. A. (2001). Acid–base interactions in adhesive modified polyethylene/metal joints; Starostina, I. A., Khasbiullin, R. R., Stoyanov, O. V., Chalykh, A. E.; Zhurnal prikladnoi khimii. 74(11), 1859–1862.
110. Burdova, E. V. (2004). Measurement of the surface-energy characteristics of synthetic rubbers and modifiers for them; Burdova, E. V., Starostina, I. A., Nefedev, E. S., Kustovsky, V. Ya. Chernov, A. V., Zaikin, A. E., Stoyanov, O. V.; "Structure and dynamics of molecular systems". Book of papers. - M. Issue 11, 150–153.
111. Kustovsky, V. Ya. (2004). The surface-energy characteristics and acidity parameters of three-layered anti-corrosion coatings; Kustovsky, V. Ya., Starostina, I. A., Khuzakhanov, R. M., Stoyanov, O. V.; "Structure and dynamics of molecular systems". Book of papers. – Moscow – Yoshkar-Ola – Ufa. Issue 11, 465–468.
112. van Oss, C. J., Good, R. J., & Chaudhury, M. K. (1986). J. Protein Chem. 5, 385–402.
113. van Oss, C. J., Chaudhury, M. K., & Good, R. J. (1987). Adv. Colloid Interface Sci. V.28, 35–60.
114. Good, R. J., & van Oss, C. J. (1991). in: Modern Approach to Wettability, Theory and Application, M. E. Schrader and, G. Loeb (Eds), Ch. 1, pp. 1–28. Plenum Press, New York.

115. Della Volpe, C., Deimichei, A., & Ricco, T. (1998). J. Adhesion Sci. Technol. V.12, 1141–1180.
116. Della Volpe, C. et al. (2003). Recent theoretical and experimental advancements in the application of the van Oss – Chaudhury – Good acid–base theory to the analysis of polymer surfaces II. Some peculiar cases; J. Adhesion Sci. Technol. 17(11), 1425–1456.
117. Della Volpe, C. et al. (2004). The solid surface free energy calculation, I. In defense of the multicomponent approach; Journal of Colloid and Interface Science. No 271, 434–453.
118. Della Volpe, C., & Siboni, S. (1997). Some reflections on acid–base solid surface free energy theories; Journal of Colloid and Interface Sci. No 195, 121–136.
119. Della Volpe, C., & Siboni, S. (2001). Troubleshooting of surface free-energy acid–base theory applied to solid surfaces: the case of Good, van Oss and Chaudhury theory; Acid–base interactions: Relevance to Adhesion Science and Technology / Editor: K. L., Mittal 2, 55–90.
120. Abraham, M. H. (1993). Chem. Soc. Rev. vol.22, 73–83.
121. Lee, L. H. (1998). J. Adhesion; v.67, 1–18.
122. Kamlet, M. J.; Abboud, J. M.; Abraham, M. H., & Taft, R. W. (1983). J. Org. Chem. V.48, 2877–2891.
123. Wu, S. (1982). Polymer Interface and Adhesion. Marcel Dekker, New York.
124. Hellwig, G. E. H., & Neumann, A. W. (1968). Proceedings of the V International Congress on Surface Active Substances, Barcelona. Sec. B. p.687.
125. Janczuk, B., & Bialopiotrowicz, T. (1990). J. Colloid Interface Sci. V.140, 362–372.
126. Neumann, A. W. & Renzow, D. (1969). Z. Phys. Chem. (Frankfurt). V. 68, P. 11.
127. Lee, L. H. (1996). Langmuir. v.12, p.1681.
128. Jan´czuk, B., Wo´jcik, W., & Zdziennicka, A. (1983). J. Colloid Interface Sci. V. 157, P. 384.
129. Starostina, I. A., Stoyanov, O. V., & Makhrova, N. V., A new approach to determining acidic and basic parameters of the surface free energy of polymers; Proceedings of the Russian Academy of Sciences – to appear.
130. Starostina, I. A., Stoyanov, O. V., Makhrova, N. V., & Nguen, D. A. (2010). Estimation of both acidic and basic parameters of the surface free energy of polymer materials; Vestnik Kazanskogo Technol. Univ. No 8, 427.
131. Bialopiotrowicz, T. (2003). Wettability of starch gel films; Food Hydrocolloids. 17, 141–147.
132. Shalel-Levanon, S., & Marmur, A. (2003). Validity and accuracy in evaluating surface tension of solids by additive approaches; Journal of Colloid and Interface Science. No 262. 489–499.
133. Walinder, M., & Gardner, D. (2002). Acid–base characterization of wood and selected thermoplastics; J. Adhesion Sci. Technol. 16(12), 1625–1649.
134. Tshabalala, M. (1997). Determination of the acid–base characteristics of lignocellulosic surfaces by Inverse Gas Chromatography; J. Appl. Pol. Sci. 65(5), 1013–1020.
135. Huang, X. et al. (2006). Surface characterization of nylon 66 by inverse gas chromatography and contact angle; Polymer Testing. 25, 970–974.
136. Santos, J. M. R., C. A., et al. (2002). Characterization of the surface Lewis acid–base properties of the components of pigmented, impact-modified, bisphenol A polycarbonate-poly(butylene terephthalate) blends by inverse gas chromatography–phase separation and phase preferences; J. Chromatogr. A. No . 969, 119–132.
137. Svendsen, J. R. et al. (2007). Adhesion between coating layers based on epoxy and silicone; Journal of Colloid and Interface Science. No 316, 678–686.

138. Hruska, Z., & Lepot, X. (2000). Ageing of the oxyfluorinated polypropylene surface: evolution of the acid–base surface characteristics with time; Journal of Fluorine Chemistry. No 105, 87–93.
139. Bismarck, A., Kumru, E., & Springer, J. (1999). Characterization of Several Polymer Surfaces by Streaming Potential and Wetting Measurements: Some Reflections on Acid–base Interactions; Journal of Colloid and Interface Science. No 217, 377–387.
140. Luner, P. E., & Oh, E. (2001). Characterization of the surface free energy of cellulose ether films; Colloids and Surfaces A: Physicochem. Eng. Aspects. No 181, 31–48.
141. Ponsonnet, L. et al. (2003). Relationship between surface properties (roughness, wettability) of titanium and titanium alloys and cell behaviour; Materials Science and Engineering. 23, 551–560.
142. Zhao, Q. et al. (2005). Surface free energies of electroless Ni–P based composite coatings; Applied Surface Science. No 240. 441–451.
143. Oh, E., & Luner, P. E. (1999). Surface free energy of ethylcellulose films and the influence of plasticizers; International Journal of Pharmaceutics. No 188, 203–219.
144. Rankl, M. et al. (2003). Surface tension properties of surface-coatings for application in biodiagnostics determined by contact angle measurements; Colloids and Surfaces B: Biointerfaces. 30, 177–186.
145. Papirer, E., Balard, H., & Vidal, A. (1988). Eur. Polym. J. No 24, 783–790.
146. Zubov, P. I., & Sukhareva, L. A. (1982). Structure and properties of polymeric coatings. M.: Khimiya, 256 p.
147. Lee, H., & Neville, K. (1967). Handbook of Epoxy Resins. McGraw-Hill, 114–118, 513–516.
148. Kestalman, V. N. (1980). Physical methods for modifying polymeric materials. M.: Khimiya, 224.
149. Kustovsky, V. Ya. (2006). Acid–base interactions and adhesion capacity in the system constituted by an epoxy coating and a metal; Kustovsky, V. Ya., Starostina, I. A., & Stoyanov, O. V.; Russian Journal of Applied Chemistry, 79(6), 930–933.
150. Kustovsky, V. Ya. (2006). Acid–base interactions and adhesion capacity in the system constituted by an epoxy coating and a metal; Kustovsky, V. Ya., Starostina, I. A., & Stoyanov, O. V.; Zhurnal prikladnoi khimii. 79(6), 940–943.
151. Kustovsky, V.Ya. (2005). The influence of acid–base interactions on the formation of adhesive joints of epoxy compositions and metals; Kustovsky, V. Ya., Starostina, I. A., & Stoyanov, O. V.; Klei. Germetiki. Tekhnologii. 12, 2–4.
152. Starostina, I. A. (2006). The influence of epoxy primer composition on its acid–base and adhesive properties; Starostina, I. A., Stoyanov, O. V., Garipov, R. M., & Kustovsky, V. Ya.; Vestnik Kazanskogo Technol. Univ. 1, 140–145.
153. Kustovsky, V.Ya. (2005). Acid–base interactions and adhesion in polymeric coating/metal systems; Kustovsky, V. Ya., Stoyanov, O. V., & Starostina, I. A.; Proceedings of young scientists and specialists. – Cheboksary: Chuvash State Univ. 138–139.
154. Kustovsky, V. Ya. (2005). Acid–base and adhesive properties of epoxy oligomer-based systems; Kustovsky, V. Ya., Stoyanov, O. V., Starostina, I. A., Garipov, R. M., & Khuzakhanov, R. M.; Proceedings of the IX Int. conference on chemistry and physicochemistry of oligomers "Oligomers-IX". Moscow, Odessa, 67.
155. Kustovsky, V. Ya. (2006). Acid–base characteristics of the modified epoxy coatings; Kustovsky, V. Ya., Stoyanov, O. V., Starostina, I. A., & Garipov, R. M.; Proceedings of the III All-Russian scientific conference "Physicochemistry of polymer processing". – Ivanovo: IGHTU. 171.

156. Starostina, I. A. (2007). The relationship between the relative acidity parameter and adhesive properties of epoxy coatings; Starostina, I. A., Stoyanov, O. V., Garipov, R. M., Zagidullin, A. I., Kustovsky, V. Ya., Koltsov, N. I., Kuzmin, M. V., Trofimov, D. M., & Petrov, V. G.; Lakokrasochnye materialy i ikh primenenie. 5, 32–36.
157. Garipov, R. M. (2007). The influence of the silicon-containing amine on the properties of epoxy coatings; Garipov, R. M., Kolpakova, M. V., Zagidullin, A. I., Stoyanov, O. V., & Starostina, I. A.; Lakokrasochnye materialy i ikh primenenie. No 7–8, 33–36.
158. Aleeva Ya. I. (2008). The influence of the nature of curing agent on the surface and adhesive characteristics of the epoxy coated metal; Aleeva, Ya. I., Starostina, I. A., & Stoyanov, O. V.; "Structure and dynamics of molecular systems". Book of papers. – Moscow – Yoshkar-Ola – Ufa. Part 3, 215–218.
159. Aleeva Ya. I. (2008). Acid–base and adhesive properties of epoxy coatings cured by complex compounds based on Lewis acids and tri(halogen)alkylphosphates; Aleeva, Ya.I., Starostina, I. A., Stoyanov, O. V., Zinoveva, E. G., Efimov, V. A., & Koltsov, N. I.; Vestnik Kazanskogo Technol. Univ. No 6, Part 1, 179–185.
160. Starostina, I. A. (2009). Acid–base interactions and their role in forecasting of polymer composites adhesion properties, Starostina, I. A., Aleeva, Y. I., Sechko, E. V., & Stoyanov, O. V., in: Encyclopedia of Polymer Composites: Properties, Performance and Applications: NY. Editors: Lechkov, M., & Prandzheva, S., 681–704
161. Garipov, R. M. (2004). Formation of flexible epoxy amine matrices upon curing and no heating – Author's abstract, Post Doctoral thesis in Chemistry. Kazan. 344 p.
162. Stoyanov, O. V. (1997). Modification of structure and properties of polyethylene coatings with polyfunctional substances; Author's abstract, Post Doctoral thesis in Engineering - Kazan. 36 p.
163. Stoyanov, O. V., Rusanova, S. N., Petukhova, O. G., & Remizov, A. B. (2001). Zhurnal prikladnoi khimii 74(7), 1774.
164. Rusanova, S. N. (2000). Modification of ethylene-vinyl acetate copolymer with saturated copolymers: Ph.D. thesis in Engineering; Rusanova, S. N.; Kazan State Technol. Univ., – Kazan
165. Starostina, I. A. (2009). The influence of acid–base properties of metals, polymers, and polymeric composite materials on adhesive interaction in metal/polymer systems; Starostina, I. A., Burdova, E. V., Sechko, E. K., Khuzakhanov, R. M., &Stoyanov, O. V.; Vestnik Kazanskogo Technol. Univ. No 3. 85–95.
166. Starostina, I. A. (2005). The role of acid–base interactions in forming adhesive polymer/metal joints; Starostina, I. A., Burdova, E. V., Kustovsky, V.Ya., & Stoyanov, O. V.; Klei. Germetiki. Tekhnologii. No 10. 16–21.
167. Burdova, E. V. (2004). Measurement of the surface-energy characteristics of synthetic rubbers and modifiers for them; Burdova, E. V., Nefedev, E. S., Starostina, I. A., Kustovsky, V. V., Chernov, A. N., Zaikin, A. E., & Stoyanov, O. V.; XI All-Russian conference "Structure and dynamics of molecular systems". Book of abstracts. Yalchik, 50.
168. Lyusova, L. R. (1987). Adhesives based on the halogen-containing polymers.; Lyusova, L. R., Polsman, G. S., Reznichenko, S. V., & Glagolev, V. A.; Subject review: Production of technical rubber goods and technical asbestos goods. M.: Khimiya, 40.
169. Chernov, A. V. (2006). Adhesive compositions for anti-corrosion insulation of pipelines with adhesive tapes working at the increased service temperature: Ph.D. thesis in Engineering; Kazan State Technol. Univ. – Kazan, 135 p.
170. Zakharchenko, P. I. (1971). A handbook for rubber producer. Zakharchenko, P. I. et al. (Ed.); M.: Khimiya. 606 p.

171. Starostina, I. A., Burdova, E. V., Khairullin, R. K., & Stoyanov, O. V. (2005). Acid–base interactions and the strength of adhesive joint in the polyethylene/butyl rubber adhesive system. Vestnik Kazanskogo Technol. Univ. No 2, Part 2, 122–125
172. Dumsky, Yu. V. (1988). Petroleum resins. M.: Khimiya. 167 p.
173. Portnoi, Ts. B., Burdova, E. V., Khairullin, R. K., Starostina, I. A., Volfson, S. I., & Stoyanov, O. V. (2006). The role of acid–base interactions in achieving adhesive strength between rubber and steel cord; Klei. Germetiki. Tekhnologii. No 6, 9–11.
174. Burdova, E. V. (2006). The influence of acid–base interactions on the adhesion of rubber mixes to brass; Burdova, E. V., Starostina, I. A., Khairullin, R. K., & Stoyanov, O. V.; Proceedings of the III All-Russian scientific conference "Physicochemistry of polymer processing". Ivanovo, 91.
175. Portnoi, Ts. B. (2005). Acid–base properties of adhesive additives for rubber mixes; Portnoi, Ts. B., Ilyasov, R. S., Khairullin, R. K., Burdova, E. V., & Starostina, I. A.; Proceedings of the VI Int. conference on intensification of petrochemical processes "Petrochemistry (2005)." Nizhnekamsk, 80.
176. Khairullin, R. K. (2005). The influence of acid–base interactions on the adhesion of rubber to the brass-coated cord; Khairullin, R. K., Portnoi, Ts. B., Ilyasov, R. S., Burdova, E. V., & Starostina, I. A.; XII All-Russian conference "Structure and dynamics of molecular systems". Book of abstracts. Yoshkar-Ola, 228.
177. Agatova, I. G. (1987). The properties of rubbers modified by the systems based on RU modifier, cobalt chelates, and alkylphenol disulfides; Agatova, I. G., Sakharova, E. V., Potapov, E. E., & Shvarts, A. G.; Kauchuk i rezina. No 11, 33–36.
178. Waulan, J. (1986). The effect of CoS /NiS on the adhesion of rubber-brass; Waulan, J., Benquian, G., Hungliang, F.; Prac. Int. Rubber Conf. "IRC 86". - Goteborg, 2, 511–513.
179. Balley, A. I., Koy, S. M. (1967). Proc. Roy. Soc. v.301, p.47.
180. Bolger, J. C. (1968). Interface Conversions for Polymer Coatings; J. C., Bolger, A. S., Michaells, Ed. P. Weiss, G. D., Cheever.; New York: Elsevier, 3.
181. Iron alloys, (1963). A short chemical encyclopedia: State scientific printing office "Sovetskaya entsiklopediya". Moscow, v.2, 21–22.
182. Finlayson, M. F., Shah, B. A. (1990). J. Adhes. Sci. and Technol. 4(5), 431–439.
183. Leggat, B. R., et al. (2002). Adhesion of epoxy to hydrotalcite conversion coatings: I. Correlation with wettability and electrokinetic measurements; Colloids and Surfaces A: Physicochem. Eng. Aspects. No . 210, 69–81.
184. Mohseni, M. et al. (2006). Adhesion performance of an epoxy clear coat on aluminum alloy in the presence of vinyl and amino-silane primers; Progress in Organic Coatings. No 57, 307–313.
185. Mykhaylyk, T. A., et al. (2003). Surface energy of ethylene-co-1-butene copolymers determined by contact angle methods; Journal of Colloid and Interface Science. No 260, 234–239.
186. Davies, D. K. (1969). Br. J. Appl. Phys. No 2, p.1533.
187. Duke, C. B., & Fabish, T. J. (1978). J. Appl. Phys. No 49, p.315.
188. McCafferty, E. (2002). Acid–base effects in polymer adhesion at metal surfaces; J. Adhesion Sci. Technol. 16(3), 239–255.
189. Starostina, I. A. (2007). The Role of Acid–base Interactions in the Formation of Polymer-Metal Adhesive Joints; Starostina, I. A., Burdova, E. V., Kustovsky, V.Ya., & Stoyanov, O. V.; Polymer Science Series, C. 49(2), 139–144.
190. Bogdanova, S. A. (2004). Some surface properties of the alternating ethylene/carbon monoxide copolymers; Bogdanova, S. A., Shashkina, O. R., Belov, G. P., Gladkov, O. N.,

Starostina, I. A., & Barabanov, V. P.; Vysokomolekulyarnye soedineniya, Series, A. 46(10), 1–7.
191. Starostina, I. A. (2001). The development of the methods for estimating the surface acid–base properties of polymeric materials; Starostina, I. A., & Stoyanov, O. V.; Vestnik Kazanskogo Technol. Univ. No 4, 58–68.
192. Starostina, I. A. (2010). Interaction of adhesives in metal-polymer systems in acid–base approach; Starostina, I. A., Khuzakhanov, R. M., Burdova, E. V., Sechko, E. K., & Stoyanov, O. V.; Polymer Science Series, D. 3(1), 26–31.
193. Garipov, R. M. (2007). The influence of the silicon-containing amine on the properties of epoxy coatings; Garipov, R. M., Kolpakova, M. V., Zagidullin, A. I., Starostina, I. A., & Stoyanov, O. V.; Lakokrasochnye materialy i ikh primenenie. No 7–8, 33–36.
194. Starostina, I. A. (2001). Acid–base Interactions at the Modified Polyethylene-Metal Interface; Starostina, I. A., Khasbiullin, R. R., Stoyanov, O. V., Chalykh, A. E.; Russian Journal of Applied Chemistry. 74(11), 1920–1923.
195. Stoyanov, O. V. (2008). Modern possibilities to estimate the acid–base properties of polymeric coatings. The review of methods and practical appendices; Stoyanov, O. V., Starostina, I. A., Burdova, E. V., & Aleeva, Ya. I.; Vestnik Kazanskogo Technol. Univ. No 5, 13–20.
196. Zav'yalova, N. B. (2008). Thermodynamic state parameters of a system as a prognostic factor of strength properties of composites with different natures of a surface; Zav'yalova, N. B., Stroganov, V. F., Stroganov, I. V., Stoyanov, O. V., Starostina, I. A., & Shayakhmetov, R. F.; Polymer Science Series, D. 1, No 2.
197. Khairullin, R. K. (2005). Acid–base properties of adhesive additives and their influence on the strength between rubber mixes and cord; Khairullin, R. K., Starostina, I. A., Ilyasov, R. S., Portnoi, Ts.B. Volfson, S. I., Burdova, E. V., & Stoyanov, O. V.; Vestnik Kazanskogo Technol. Univ. No 2 Part 2, 107–115.
198. Zavyalova, N. B. (2007). Thermodynamic parameters of the state of a system as a factor of forecasting strength properties of composite materials having different kinds of surfaces; Zavyalova, N. B., Stroganov, V. F., Stroganov, I. V., Stoyanov, O. V., Starostina, I. A., & Shayakhmetov, R. F.; Klei. Germetiki. Tekhnologii. No 11, 20–29.
199. Starostina, I. A. (2009). Adhesive interaction in metal/polymer systems in view of the acid–base approach; Starostina, I. A., Khuzakhanov, P. M., Burdova, E. V., Sechko, E. K., & Stoyanov, O. V.; Klei. Germetiki. Tekhnologii. No 7, 11–18.
200. Khairullin, R. K. (2006). Estimation of the adhesion of the components of the combined adhesive tape in overlap in view of acid–base interaction. Khairullin, R. K., Burdova, E. V., Starostina, I. A., & Stoyanov, O. V.; Klei. Germetiki. Tekhnologii. No 6, 28–30.
201. Stroganov, V. F. (2010). Epoxy polymer adhesive primers for anti-corrosion pipeline insulation; Stroganov, V. F., Stroganov, I. V., Akhmetchin, A. S., Stoyanov, O. V., & Starostina, I. A.; Izvestiya Kazanskogo Architekturno-Stroitelnogo Univ. No 1, 342–346.
202. Starostina, I. A. (2008). Quantitative characteristic of acid–base properties of polymeric coatings in adhesive joints; Starostina, I. A., Burdova, E. V., Kurnosov, B., & Stoyanov, O. V.; All materials. Encyclopedic reference book. 6, 16–20.
203. Aleeva Ya. I. (2008). The influence of curing agent nature on the surface and adhesive characteristics of epoxy coatings applied on metal; Aleeva, Ya. I., Starostina, I. A., & Stoyanov, O. V.; "Structure and dynamics of molecular systems": Book of papers. – Moscow – Yoshkar-Ola – Ufa. Part 3, 215–218.
204. Starostina, I. A. (2002). Polymeric powdered composition for coatings. Patent for invention; Starostina, I. A., Stoyanov, O. V., Chalykh, A. E., Khasbiullin, R. R., & Deberdeev, R.Ya.; No 2186, 79

205. Bogdanova, S. A. (1997). Determination of the surface free energy of solids (methodical guidelines); Bogdanova, S. A., Starostina, I. A., Stoyanov, O. V., & Potapova, M. V.; Kazan State Technol. Univ. Kazan. 12 p.
206. Bogdanova, S. A., Shashkina, O. R., I. A., Starostina, Stoyanov, O. V., & Belov, G. P. (2002). Wetting of polyolefine low-energy surfaces with nonionic surfactants. XII Int. conference "Surface forces. Deryagin reedings" – Zvenigorod. 150, 2.
207. Starostina, I. A. (2003). Acid–base and energy characteristics of organic and inorganic surfaces; Starostina, I. A., Bogdanova, S. A., & Stoyanov, O. V., XVII Mendeleev conference on general and applied chemistry "Achievements and prospects of chemical science" – Kazan. – poster session (L-Ya), 282.
208. Starostina, I. A., Stoyanov, O. V., Chalykh, A. E., Khasbiullin, R. R. (2000). Acid–based Interaction in the polyethylene-steel systems. Int. conference "Organometallic compounds – materials of the future millenium" – Nizh. Novgorod. 146.
209. Shashkina, O. R., Bogdanova, S. A., Belov, G. P., Golodkov, O. N., Starostina, I. A., & Barabanov, V. P. (2000). The surface free energy and acid–base properties of ethylene/carbon monoxide copolymers; II All-Russian Kargin symposium "Chemistry and physics of polymers in the early XXI". – Chernogolovka. 84.
210. Starostina, I. A., Stoyanov, O. V., Deberdeev, R. Ya, & Mukmeneva, N. A. (1997). Int. conference "Fundamental problems of polymer science" - M. 74.
211. Khasbiullin, R. R., Starostina, I. A., & Stoyanov, O. V. (1996). Acid–base interactions in the modified HDPE/steel system / VIII Int. conference of young scientists "Synthesis, investigation of properties, and processing of high molecular weight compounds" – Kazan, 98.
212. Shapoval, G. S., Bagry, V. A. et al. (1985). Study on cathodic delamination of polymeric coatings; Zhurnal prikladnoi khimii. No 11, 2562–2565.
213. Kusaka, I., & Suetaka, W. (1980). Spectrochim.Acta, 36A–647.
214. Shevtsova, K. V., Baranova, T. V., Pitsyk, V. A., & Vashchenko, Yu. N. (2004). On the mechanism of the promoting action of the blocked polyisocyanates and composite modifiers made thereof in adhesive rubber/brass system; Voprosy khimii i khimicheskoi tekhnologii. No 1, 151–154.
215. Shmurak, I. L., & Matyukhin, S. A. (1989). The bonding strength in the brass-coated steel cord/rubber system and the ways of its enhancement by modifying steel cord surface. M.: TsNIITEneftekhim, 93 p.
216. Agayants, L. L., Lykin, V. L., Maloenko, V. L., & Shvarts, A. G. (1982). Modification of rubbers for enhancing the strength of adhesion to metals; M.: TsNIITEneftekhim, 84 p.

INDEX

B

C

D

E

F

G